CAMBRIDGE International Examinations

International Edition for IGCSE and O
Biology
Mary Jones and Geoff Jones

CAMBRIDGE
UNIVERSITY PRESS

PUBLISHED BY THE PRESS SYNDICATE OF THE UNIVERSITY OF CAMBRIDGE
The Pitt Building, Trumpington Street, Cambridge, United Kingdom

CAMBRIDGE UNIVERSITY PRESS
The Edinburgh Building, Cambridge CB2 2RU, UK
40 West 20th Street, New York, NY 10011-4211, USA
477 Williamstown Road, Port Melbourne, VIC 3207, Australia
Ruiz de Alarcón 13, 28014 Madrid, Spain
Dock House, The Waterfront, Cape Town 8001, South Africa

http://www.cambridge.org

© Cambridge University Press 2002

This book is in copyright. Subject to statutory exception and to the provisions of relevant collective licensing agreements, no reproduction of any part may take place without the written permission of Cambridge University Press.

First published 2002

Prepared for publication by Geoff Jones

Printed in the United Kingdom at the University Press

Typeface Meridien   *System* QuarkXPress®

A catalogue record for this book is available from the British Library

ISBN 0 521 89117 5   paperback

ACKNOWLEDGEMENTS

We are grateful to the following for permission to reproduce photographs:

2l, Bruce Iverson/SPL; 2r, 35, 60, 65, 119, 134, Biophoto Associates; 36, Andrew Syred/SPL; 52, Touchline Photos/Mike Powell; 76, Dr RHF Hunter; 83, Mrs J M Bebb; 120, Camilla Jessel; 128, Wendy Lee; 129t, courtesy of the National Portrait Gallery, London; 129b, Lee Rue/Frank Lane Photographic Agency; 130, Stephen Dalton/NHPA; 132, NHPA/Agence Nature; 133, John Kaddu; 137, CNRI/SPL; 138, Ron Gilling/Panos Pictures; 142, Philippe Plailly/SPL; 145, David Hoffman; 150, 152, C James Webb; 153, Fergus Smith/Life File; 155t, Abbie Enock/Travel Ink/Life File; 155b, 177 Michael Brooke; 173l, Sean Sprague/Panos Pictures; 173r, Martin Bond/SPL; 175l, Michael Harvey/Panos Pictures; 175r, Mary Jones

The line illustrations in this book are © Geoff Jones 1984, 1987, 1995, 2002

Cover image of DNA Chain courtesy of Getty Images – Michel Tcherevkoff

Past examinations questions are reproduced by permission of the University of Cambridge Local Examinations Syndicate.

Contents

	Introduction	iv				
1	**Cells**	1	8	**Reproduction**	71	
	Cell structure	1		Asexual reproduction	71	
	Cells and organisms	3		Sexual reproduction	73	
2	**Diffusion, osmosis and active transport**	6		Sexual reproduction in a mammal	74	
3	**Enzymes**	12		Sexual reproduction in flowering plants	82	
	Making use of enzymes	15		Sexual and asexual reproduction	89	
4	**How animals feed**	17	9	**Coordination and response**	92	
	Carbohydrates	17		The eye	92	
	Proteins	19		Coordination	95	
	Fats	20		Nervous systems	95	
	Balanced diet	22		The endocrine system	100	
	Digestion	24		Coordination in plants	102	
	Digestion in humans – teeth	26	10	**Homeostasis and excretion**	106	
	Digestion in humans – the alimentary canal	29		The control of body temperature	106	
5	**How green plants feed**	34		The control of blood glucose content	110	
	Leaves	34		Excretion	112	
	Photosynthesis experiments	39		The human excretory system	113	
6	**Respiration**	44	11	**Support and movement**	116	
	Gaseous exchange	46		The human skeleton	116	
	Gaseous exchange in humans	46		Support in plants	118	
	Smoking	52	12	**Inheritance and evolution**	119	
7	**Transport**	54		Inheritance	122	
	Layout of a mammal transport system	54		Variation and selection	126	
	Blood vessels	57		Natural selection	129	
	Blood	59	13	**Health, disease and medicine**	134	
	Lymph and tissue fluid	62		Genetic diseases	134	
	Transport in a flowering plant	64		Infectious diseases	135	
	Transport of water	65		Body defences against infectious diseases	139	
	Uptake of minerals	70		Life-style and disease	143	
	Transport of manufactured food	70				

14	**Making use of microorganisms**	**147**
	Growing microorganisms	147
	Sewage treatment	150
	Making food with microorganisms	152
	Microorganisms and medicine	155
	Genetic engineering	156
15	**Living organisms in their environment**	**159**
	Studying ecosystems	160
	Food and energy in an ecosystem	160
	Nutrient cycles	163
	Population size	165
16	**Humans and the environment**	**170**
	Global warming	171
	Acid rain	173
	Deforestation	175
	Water pollution	176
	Pesticides	178
	Conservation	179
17	**The diversity of life**	**181**
	Kingdom Bacteria	182
	Kingdom Fungi	182
	Kingdom Plantae	183
	Kingdom Animalia	183

Apparatus required for practicals	**186**
Past examination questions	**189**
Glossary	**199**
Index	**204**

Introduction

This book has been written to help you to do well in your Cambridge International Examinations 'O' level or IGCSE Biology examination. We hope that you enjoy using it.

'O' level and IGCSE material

Most of the book contains material that you will need whichever of these two examinations you will be taking, but:
- anything that you need to know for 'O' level, but *not* for IGCSE, is marked with stripes like this;
- anything that you need to know for IGCSE but *not* for 'O' level is marked with a bar like this.

You may still like to read these parts of the book, even if you do not need the material for your examination, as they can help you to gain a wider understanding of the topics.

Assessment Objectives

Whichever examination you are studying for, you will be tested on three sets of Assessment Objectives.

The first of these is **Knowledge with Understanding**. You are expected to know and understand all of the facts and concepts listed in the syllabus. These are all fully covered in this book. When you are revising for a test or for an examination, make sure that you *understand* these facts and concepts. This will make it much easier for you to learn and remember them.

The second set of Assessment Objectives is **Handling Information and Solving Problems**. Questions testing this will expect you to use your knowledge and understanding in an unfamiliar context. The examiners will set questions that look unfamiliar to you. However, if you have a good knowledge and understanding of the facts and concepts covered in this book, you should be able to tackle such questions confidently, applying your knowledge to these new situations.

The third set of Assessment Objectives is **Experimental Skills and Investigations**. This involves practical work, and your examination will test you on your practical skills. So it is very important that you work hard on improving these skills. If at all possible, you should try to do plenty of practical work. There are many ideas for this in this book.

Sometimes, there are facts you need to know that are deliberately not explained in the book, but which you will find out for yourself by carrying out an experiment. Sometimes, your syllabus expects you to know how to carry out a particular experiment – check this out in the latest version of your syllabus.

Definitions

Biology is full of technical terms! It is very important that you use these terms precisely, and understand their exact meanings. When a new term is first introduced, it is shown in **bold** and its meaning is explained. You can also find definitions in the Glossary, on pages 199 to 203.

Questions

There are many questions within each Chapter, and also some at the ends of each Chapter. Most of these questions are to help you to think about what you have read, and to encourage your brain to store what you know so that you will find it easier to remember it. The more times you think about something, the more likely your brain is to store it in your long-term memory!

At the end of the book, on pages 189 to 198, you will find questions from past CIE 'O' level and IGCSE examination papers. These are arranged in approximately the same order as the Chapters in the book. However, quite a few of them cover material from two or more different Chapters – these are especially good practice for you, because they help you to think about things in a slightly different way, which can increase your understanding.

Each question states whether it is from an 'O' level (5090) or from an IGCSE (0610) examination paper. However, most of them are suitable for you to use, whichever of these two examinations you will be taking. (There are also a few questions from a different paper, 5096, which are also good practice for you.) If part of a question tests something that is only needed for 'O' level, or only needed for IGCSE, then it is marked with grey stripes or a grey bar. Do remember, too, that syllabuses change. So, for example, you may not need to know something that was set in an examination two years ago. Always check with your syllabus to find out exactly what you need and what you do *not* need to know.

1 Cells

1.1 All living things have certain characteristics.

Biology is the study of living things, which are often called **organisms**. Living organisms have several features or **characteristics** which make them different from objects which are not alive.

1. They **reproduce**.
2. They take in nutrients. This is known as feeding or **nutrition**.
3. They **respire** – that is, they release energy from their food, often by combining it with oxygen.
4. They **grow**.
5. They **excrete** – that is, they get rid of waste substances that have been made by chemical reactions, known as **metabolic reactions**, going on inside them.
6. They **move**.
7. They are **sensitive** – that is, they can sense and respond to changes in their surroundings.
8. They are made of **cells**.

Cell structure

1.2 Microscopes are used to study cells.

All living things are made of cells. Cells are very small, so large organisms contain millions of cells.

To see cells clearly, you need to use a microscope. The kind of microscope used in a school laboratory is called a light microscope because it shines light through the piece of animal or plant you are looking at. It uses glass lenses to magnify and focus the image. A very good light microscope can magnify about 1500 times, and can show all the structures in Figs 1.1 and 1.2.

> **Questions**
> 1. List the characteristics of living things.
> 2. How many times can a good light microscope magnify?

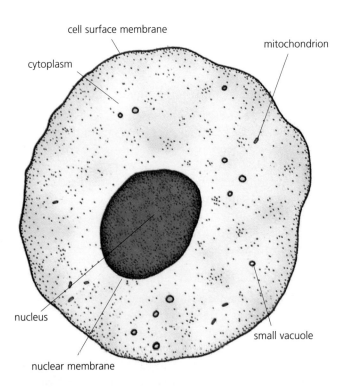

1.1 A typical animal cell as seen with a light microscope. This is a liver cell.

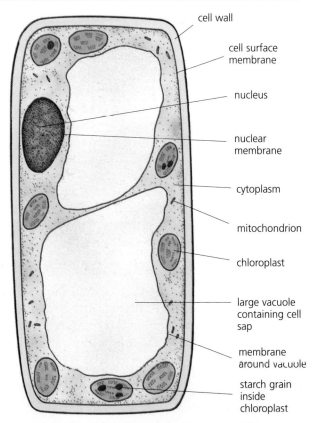

1.2 A typical plant cell as seen with a light microscope. This is a palisade cell.

1.3 All cells have a cell membrane.

Whatever sort of animal or plant they come from, all cells have a **cell membrane** around the outside. Inside the cell membrane is a jelly-like substance called **cytoplasm**, in which are found many small structures called **organelles**. The most obvious of these is usually the **nucleus** (Fig 1.3).

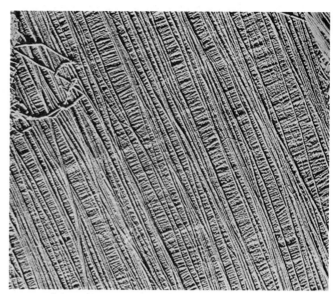

1.4 Cellulose fibres in a plant cell wall.

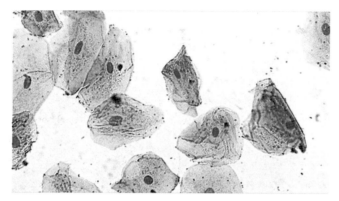

1.3 Cheek cells seen through a light microscope. The nuclei are stained a deeper colour than the cytoplasm. Cheek cells are very thin and some of the cells in this photograph have been accidentally folded during the preparation of the specimen.

1.4 Plant cells have a cell wall.

All plant cells are surrounded by a **cell wall** made of **cellulose**. Paper, which is made from cell walls, is also made of cellulose. Animal cells never have cell walls. Cellulose belongs to a group of substances called **polysaccharides**, which are described in Section 4.7. Cellulose forms fibres which criss-cross over one another to form a very strong covering to the cell (Fig 1.4). This helps to protect and support the cell. It stops the plant cell from bursting when it absorbs water by osmosis. This is described in Section 2.7.

Because of the spaces between the fibres, even very large molecules are able to go through the cellulose cell wall. It is therefore said to be **fully permeable**.

1.5 Cell membranes are partially permeable.

All cells have a membrane surrounding the cell. It is called the **cell surface membrane** or **plasma membrane**. In a plant cell, it is very difficult to see, because it is right against the cell wall.

The cell surface membrane is a very thin layer of protein and fat. It is very important to the cell because it controls what goes in and out of it. It is said to be **partially permeable**, which means that it will let some substances through but not others.

1.6 Cytoplasm is a complex solution.

Cytoplasm is a clear jelly. It is nearly all water; about 70% is water in many cells. It contains many substances dissolved in it, especially proteins. Many metabolic reactions take place in the cytoplasm.

1.7 Most cells contain vacuoles.

A vacuole is a space in a cell, surrounded by a membrane, and containing a solution. Most plant cells have very large vacuoles, which contain a solution of sugars and other substances called **cell sap**.

Animal cells have much smaller vacuoles, which may contain food or water.

1.8 Chloroplasts trap the energy of sunlight.

Chloroplasts are never found in animal cells, but most of the cells in the green parts of plants have them. They contain the green colouring or pigment called **chlorophyll**. Chlorophyll absorbs sunlight, and the energy of sunlight is then used for making food for the plant by **photosynthesis** (Chapter 5).

Chloroplasts often contain starch grains, which have been made by photosynthesis. Animal cells never contain starch grains.

1.9 Mitochondria release energy from food.

Most cells have mitochondria, because it is here that the cell releases energy from food. The energy is needed to help the cell move and grow. Mitochondria are sometimes called the 'powerhouses' of the cell. The energy is released by combining food with oxygen, in a process called **aerobic respiration**. The more active a cell, the more mitochondria it has.

1.10 The nucleus stores inherited information.

The nucleus is where the information is stored which helps the cell to make the right sort of proteins. The information is kept on the **chromosomes**, which are inherited from the organism's parents.

Chromosomes are very long, but so thin that they cannot easily be seen even with a microscope. However, when the cell is dividing, they become short and thick, and can be seen with a good light microscope.

Questions

1. What sort of cells are surrounded by a cell surface membrane?
2. What are plant cell walls made of?
3. What does fully permeable mean?
4. What does partially permeable mean?
5. What is the main constituent of cytoplasm?
6. What is a vacuole?
7. What is cell sap?
8. Chloroplasts contain chlorophyll. What does chlorophyll do?
9. What happens inside mitochondria?
10. What is stored in the nucleus?
11. Why can chromosomes be seen only when a cell is dividing?

Cells and organisms

1.11 Similar cells are grouped to form tissues.

A large organism such as yourself may contain many millions of cells, but not all the cells are alike. Almost all of them can carry out most of the activities which are characteristic of living things, but many of them specialise in doing some of these better than other cells do. Muscle cells, for example, are specially adapted for movement. Most cells in the leaf of a plant are specially adapted for making food by photosynthesis.

Often, cells which specialise in the same activity will be found together. A group of cells like this is called a **tissue**. An example of a tissue is a layer of cells lining your stomach. These cells make enzymes to help to digest your food (Fig 1.6, overleaf).

The stomach also contains other tissues. For example, there is a layer of muscle in the stomach wall, made of cells which can move. This muscle tissue makes the wall of the stomach move in and out, churning the food and mixing it up with the enzymes.

Practical 1.1 Looking at animal cells

The easiest place to find animal cells is on yourself. If you colour or stain the cells, they are quite easy to see using a light microscope (Fig 1.5).

1. Using a clean fingernail or section lifter gently rub off a little of the lining from the inside of your cheek.
2. Put your cells onto the middle of a clean microscope slide, and gently spread them out. You will probably not be able to see anything at all at this stage.
3. Put on a few drops of methylene blue.
4. Gently lower a coverslip over the stained cells, trying not to trap any air bubbles.
5. Use filter paper or blotting paper to clean up the slide, and then look at it under the low power of a microscope.
6. Make a labelled drawing of a few cells.

Questions

1. Which part of the cell stained the darkest blue?
2. Is the cell surface membrane permeable or impermeable to methylene blue? Explain your answer.

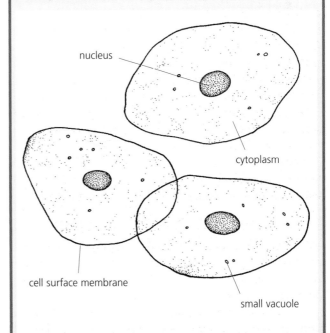

1.5 Cheek cells seen through a light microscope.

1.12 Tissues make up organs and systems.

All tissues in the stomach work together, although they each have their own job to do. A group of tissues like this makes up an **organ**. The stomach is an organ. Other organs include the heart, the kidneys and the lungs.

The stomach is only one of the organs which help in the digestion of food. The mouth, the intestines and the stomach are all part of the digestive system. The heart is part of the circulatory system, while each kidney is part of the excretory system.

The way in which organisms are built up can be summarised like this:

Organelles make up **cells,** which make up **tissues**, which make up **organs**, which make up **systems**, which make up **organisms**.

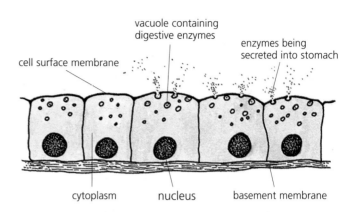

1.6 Cells lining the stomach; an example of a tissue.

Practical 1.2 Looking at plant cells

To be able to see cells clearly under a microscope, you need a very thin layer. It is best if it is only one cell thick. An easy place to find such a layer is inside an onion bulb (Fig 1.7).

1. Cut a small piece from an onion bulb, and use forceps to peel a small piece of thin skin, called epidermis, from the inside of it. Do not let it get dry.
2. Put a drop or two of water onto the centre of a clean microscope slide. Put the piece of epidermis into it, and spread it flat.
3. Gently lower a coverslip onto it.
4. Use filter paper or blotting paper to clean up the slide, and then look at it under the low power of a microscope.
5. Make a labelled drawing of a few cells.
6. Using a pipette, take up a small amount of iodine solution. Very carefully place some iodine solution next to the edge of the coverslip. The iodine solution will seep underneath the coverslip. To help it do this, you can place a small piece of filter paper next to the opposite side of the coverslip, which will soak up some of the liquid.
7. Look at the slide under the low power of the microscope. Note any differences between what you can see now, and what it looked like before adding the iodine solution.

Questions

1. Name two structures which you can see in these cells, but which you could *not* see in the cheek cells.
2. Most plant cells have chloroplasts, but these onion cells do not. Suggest a reason for this.
3. Explain the change in appearance of the cells when you stained them with iodine solution.
4. What would cheek cells look like if you stained them with iodine solution?

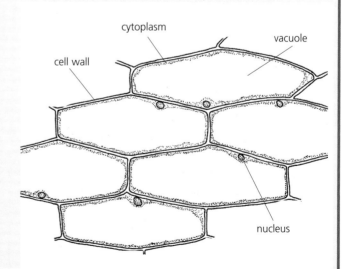

1.7 Onion epidermis cells seen through a light microscope.

1.13 Cell structure is matched to function.

We have seen that different cells are specialised to carry out different functions. In most cases, they have a particular structure that helps them to carry out these functions efficiently.

You will meet many examples of specialised cells later in this book. Some particularly good examples include:

- Root hair cells of plants, which are specialised for absorbing water and inorganic ions from the soil into the plant. These are described on page 65.
- Xylem vessel elements, which are specialised for transporting water up through a plant from its roots into its leaves, and also for helping to support the plant. These are described on page 64.
- Red blood cells of animals, which are specialised for transporting oxygen from the lungs to all the other parts of the body. These are described on page 59.
- Ciliated cells lining the air passages of animals, which are specialised for sweeping mucus, in which bacteria and dust have been trapped, up into the back of the throat. These are described on page 47.
- Muscle cells, which can contract and cause movement. These are described on page 117.

Table 1.1 A comparison between animal and plant cells.

Similarities

1. Both have a cell surface membrane surrounding the cell.
2. Both have cytoplasm.
3. Both contain a nucleus.
4. Both contain mitochondria.

Differences

	Plant cells	Animal cells
1	Have a cellulose cell wall outside the cell surface membrane	Have no cell wall
2	Often have chloroplasts containing chlorophyll	Have no chloroplasts
3	Often have very large vacuoles, containing cell sap	Have only small vacuoles
4	Often have starch granules	Never have starch granules; sometimes glycogen granules
5	Often regular in shape	Often irregular in shape

Chapter revision questions

1. Arrange these structures in order of size, beginning with the smallest.

   stomach, mitochondrion, starch grain, cheek cell, nucleus

2. For each of the following, state whether it is an organelle, a cell, a tissue, an organ, a system, or an organism.

   (a) heart (b) chloroplast (c) nucleus (d) cheek cell (e) onion epidermis (f) onion bulb (g) onion plant (h) mitochondrion (i) human being (j) lung

3. State which part of a plant cell:

   (a) makes food by photosynthesis
   (b) releases energy from food
   (c) controls what goes in and out of the cell
   (d) stores information about making proteins
   (e) contains cell sap
   (f) protects the outside of the cell.

4. A tree is a living organism, but a bicycle is not. Using these as examples, explain the differences between living and non-living things.

5. With the aid of large labelled diagrams, make a comparison of a typical plant cell and a typical animal cell. Whenever possible, explain the reasons for their differences and similarities.

6. Use the index to find out about the structure and functions of each of the following kinds of cells. For each one, explain how its structure helps it carry out its function.

   palisade cell, root hair cell, guard cell, xylem cell, phloem sieve cell, sperm cell, egg cell, red blood cell

7. Use the index to find out about each of the following systems in humans. For each system:

   (a) list the main functions of the system
   (b) outline the main functions of the system.

   nervous system, circulation (blood) system, breathing (gas exchange) system, urinary system, male and female reproductive systems

2 Diffusion, osmosis and active transport

2.1 Diffusion results from random movement.

Atoms, molecules and ions are always moving. The higher the temperature, the faster they move. In a solid substance the particles are held close together by forces between them, so they just vibrate on the spot. In a liquid they can move more freely, knocking into one another and rebounding. In a gas they are freer still. Molecules and ions can also move freely when they are in solution.

When they can move freely, particles tend to spread themselves out as evenly as they can. Imagine, for example, a rotten egg in one corner of a room, giving off hydrogen sulphide gas. To begin with, there will be a very high concentration of the gas near the egg, but none in the rest of the room. However, before long the hydrogen sulphide molecules have spread throughout the air in the room. Soon, you will not be able to tell where the smell first came from – the whole room will smell of hydrogen sulphide.

The hydrogen sulphide molecules have spread out or **diffused** through the air. Diffusion is the net movement of particles from a place where they are in a high concentration, to a place where they are in a lower concentration – that is, down a concentration gradient. Diffusion evens out the distribution of particles.

2.2 Diffusion is important to organisms.

Organisms obtain many of their requirements, and get rid of many of their waste products, by diffusion. Plants need carbon dioxide for photosynthesis. This diffuses from the air into the leaves, through the stomata. It does this because there is a lower concentration of carbon dioxide inside the leaf, where the cells are using it up. Outside the leaf in the air, there is a higher concentration. Carbon dioxide molecules therefore diffuse into the leaf, down the concentration gradient.

Oxygen, which is a waste product of photosynthesis, diffuses out in the same way. There is a higher concentration of oxygen inside the leaf, because it is being made there. Oxygen therefore diffuses out through the stomata into the air.

Diffusion is also important in gas exchange for respiration in animals and plants. In mammals, some of the products of digestion are absorbed from the ileum by diffusion.

Practical 2.1 To show diffusion in a solution

1. Fill a gas jar with water. Leave it for several hours to let it become completely still.
2. Drop a small crystal of potassium permanganate into the water.
3. Make a labelled drawing of the gas jar to show how the colour is distributed.
4. Leave the gas jar completely undisturbed for several days.
5. Make a second drawing to show how the colour is distributed.

You can try this with other salts as well, such as copper sulphate or potassium dichromate.

Questions

1. Why was it important to leave the water to become completely still before the crystal was put in?
2. Explain how colour spread through the water.
3. If the water had been warmer, diffusion would have happened more quickly. Using your knowledge of how temperature affects the movement of particles, explain why this is so.

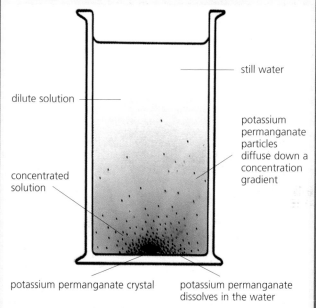

2.1 Diffusion of potassium permanganate.

Questions
1. What effect does temperature have on the movement of atoms, molecules and ions?
2. What is diffusion?
3. List three examples of diffusion which occur in living organisms.

2.3 In osmosis, water diffuses through a partially permeable membrane.

Fig 2.2 illustrates a concentrated sugar solution, separated from a dilute sugar solution by a membrane. The membrane has holes or pores in it, which are extremely small. An example of a membrane like this is Visking tubing.

Water molecules are very small. Each one is made of two hydrogen atoms and one oxygen atom. Sugar molecules are many times larger than this.

In Visking tubing, the holes are big enough to let the water molecules through, but not the sugar molecules. It is called a **partially permeable membrane** because it will let some molecules through but not others.

There is a higher concentration of sugar molecules on the right-hand side of the membrane in Fig 2.2, than on the left-hand side. If the membrane was not there, the sugar molecules would diffuse from the concentrated solution into the dilute one until they were evenly spread out. However, they cannot do this because the pores in the membrane are too small for them to get through.

There is also a concentration gradient for the water molecules. On the left-hand side of the membrane, there is a high concentration of water molecules. On the right-hand side, the concentration of water molecules is lower because a lot of the space is taken up by sugar molecules. The water molecules therefore diffuse from the left-hand side into the right-hand side. They can do this because the pores in the membrane are large enough for them to get through.

What is the result of this? Water has diffused from the dilute solution, through the partially permeable membrane, into the concentrated solution. The concentrated solution will become more dilute, because of the extra water molecules coming into it.

This process is called **osmosis**. Osmosis is the diffusion of water molecules from a place where they are in a higher concentration (such as a dilute sugar solution), to a place where the water molecules are in a lower concentration (such as a concentrated sugar solution) through a partially permeable membrane.

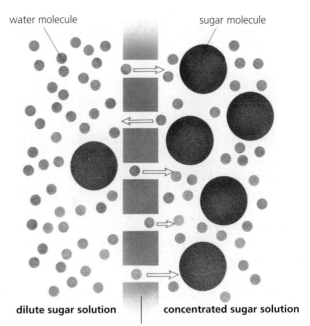

There are more water molecules on the left-hand side of the membrane than on the right-hand side, so there is a concentration gradient for water molecules. They travel down the gradient, from left to right.

Is there a concentration gradient for sugar molecules? Why don't they travel down it?

2.2 Osmosis

It is actually rather confusing to talk about the 'concentration' of water molecules, because the term 'concentration' is normally used to mean the concentration of the solute dissolved in the water. It is much better to use a different term instead. We say that a dilute solution (where there is a lot of water) has a **high water potential**. A concentrated solution (where there is less water) has a **low water potential**. In Fig 2.2, there is a high water potential on the left-hand side, and a low water potential on the right-hand side. There is a **water potential gradient** between the two sides. The water molecules diffuse down this water potential gradient, from a high water potential to a low water potential.

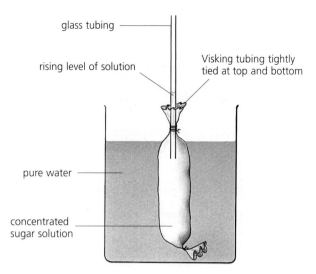

Water diffuses from the pure water into the sugar solution, down its concentration gradient. Sugar molecules cannot diffuse out, because the pores in the Visking tubing are too small for them. Therefore, the level of the sugar solution rises, as it is diluted by the water diffusing into it by osmosis.

2.3 Apparatus to demonstrate osmosis.

2.4 Cell membranes are partially permeable.

Cell membranes behave very like Visking tubing. They will let some substances pass through them, but not others. They are partially permeable membranes.

There is always cytoplasm on one side of any cell membrane. Cytoplasm is a solution of proteins and other substances in water. There is usually a solution on the other side of the membrane, too. Inside large animals, cells are surrounded by tissue fluid (Section 7.18). In the soil, the roots of plants are often surrounded by a film of water.

So, cell membranes separate two different solutions – the cytoplasm, and the solution around the cell. If the solutions are of different concentrations (that is, different water potentials), then osmosis will occur.

Questions

1 Which is larger – a water molecule or a sugar molecule?
2 What is meant by a partially permeable membrane?
3 Give two examples of partially permeable membranes.
4 How would you describe a solution which has a high concentration of water molecules?
5 What is osmosis?
6 In Fig 2.3, which liquid has a relatively high water potential, and which has a relatively low water potential?

2.5 Animal cells burst in pure water.

Fig 2.4 illustrates an animal cell in pure water. The cytoplasm inside the cell is a fairly concentrated solution. The proteins and many other substances dissolved in it are too large to get through the cell membrane. Water molecules, though, can get through.

If you compare this situation with Fig 2.2, you will see that they are similar. The dilute solution in Fig 2.2 and the pure water in Fig 2.4 are each separated from a concentrated solution by a partially permeable membrane. In Fig 2.4 the concentrated solution is the cytoplasm and the partially permeable membrane is the cell membrane. Therefore osmosis will occur. Water molecules will diffuse from the dilute solution into the concentrated solution. What happens to the cell? As more and more water enters it, it swells. The cell membrane has to stretch as the cell gets bigger, until eventually the strain is too much, and the cell bursts.

2.6 Animal cells shrink in concentrated solutions.

Fig 2.5 illustrates an animal cell in a concentrated solution. If this solution is more concentrated than the cytoplasm, then the water molecules will diffuse out of the cell. Look at Fig 2.2 to see why.

As the water molecules go out through the cell membrane, the cytoplasm shrinks. The cell shrivels up.

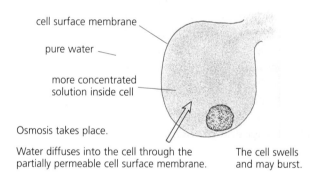

Osmosis takes place.
Water diffuses into the cell through the partially permeable cell surface membrane. The cell swells and may burst.

2.4 An animal cell in pure water.

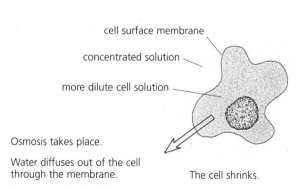

Osmosis takes place.
Water diffuses out of the cell through the membrane. The cell shrinks.

2.5 An animal cell in a concentrated solution.

2.7 Plant cells do not burst in pure water.

Fig 2.6 illustrates a plant cell in pure water. Plant cells are surrounded by a cell wall. This is fully permeable, which means it will let any molecules go through it, so osmosis will not occur across it.

Although it is not easy to see, a plant cell also has a cell membrane just like an animal cell. The cell membrane is partially permeable. A plant cell in pure water will take in water by osmosis through its partially permeable cell membrane in the same way as an animal cell. As the water goes in, the cytoplasm and vacuole will swell.

However, the plant cell has a very strong cell wall around it. The cell wall is much stronger than the cell membrane and it stops the plant cell from bursting. The cytoplasm presses out against the cell wall, but the cell wall resists and presses back on the contents.

A plant cell in this state is rather like a blown-up tyre – tight and firm. It is said to be **turgid**. The turgidity of its cells helps a plant that has no wood in it to stay upright, and keeps the leaves firm. Plant cells are usually turgid.

2.8 Plant cells plasmolyse in concentrated solutions.

Fig 2.7 illustrates a plant cell in a concentrated solution. Like the animal cell in Fig 2.5, it will lose water by osmosis. The cytoplasm and the vacuole will shrink. As the cytoplasm shrinks, it stops pushing outwards on the cell wall. Like a tyre when some of the air has leaked out, the cell becomes floppy. It is said to be **flaccid**. If the cells in a plant become flaccid, then the plant loses its firmness and begins to wilt.

If the solution is *very* concentrated, then a lot of water will diffuse out of the cell. The cytoplasm and vacuole go on shrinking. The cell wall, though, is too stiff to be able to shrink much. As the cytoplasm shrinks further and further into the centre of the cell, the cell wall gets left behind. The cell membrane, surrounding the cytoplasm, tears away from the cell wall (Fig 2.7).

A cell like this is said to be **plasmolysed**. This does not normally happen because plant cells are not usually surrounded by very strong solutions. However, you can make cells become plasmolysed if you do Practical 2.2. Plasmolysis usually kills a plant cell because the cell membrane is damaged as it tears away from the cell wall.

Osmosis is very important to plants, because it is the way in which water is absorbed into their roots. This is described in Section 7.27.

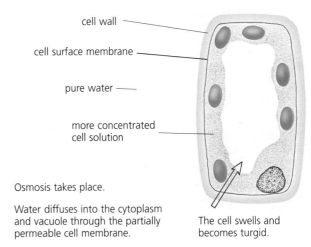

Osmosis takes place.

Water diffuses into the cytoplasm and vacuole through the partially permeable cell membrane.

The cell swells and becomes turgid.

2.6 A plant cell in pure water.

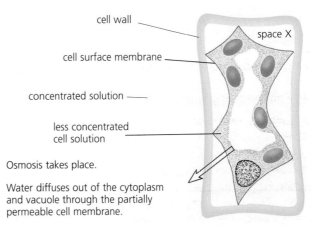

Osmosis takes place.

Water diffuses out of the cytoplasm and vacuole through the partially permeable cell membrane.

First, the cell shrinks slightly, and becomes flaccid; then the cell membrane pulls away from the cell wall, and the cell is plasmolysed.

2.7 A plant cell in a concentrated solution.

Questions

1. What happens to an animal cell in pure water?
2. Explain why this does not happen to a plant cell in pure water.
3. Which part of a plant cell is (a) fully permeable and (b) partially permeable?
4. What is meant by a turgid cell?
5. What is plasmolysis?
6. How can plasmolysis be brought about?
7. In Fig 2.7, what fills space X? Explain your answer.
8. Describe the events shown in Fig 2.6 and 2.7 in terms of water potential.

2.9 Cells take in substances by active transport.

There are many occasions when cells need to take in substances which are only present in small quantities around them.

Root hair cells, for example, take in nitrate ions from the soil.

Very often, the concentration of nitrate ions inside the root hair cell is higher than the concentration in the soil. The diffusion gradient for the nitrate ions is out of the root hair, and into the soil.

Despite this, the root hair cells are still able to take nitrate ions in. They do it by a process called **active transport**. Active transport is an energy-consuming process by which substances are transported against a concentration gradient. Fig 2.8 shows how active transport takes place.

In the cell membrane of the root hair cells are special **carrier proteins**. Carrier proteins pick up nitrate ions from outside the cell, and then change shape in such a way that they push the nitrate ions through the cell membrane and into the cytoplasm of the cell.

As its name suggests, active transport uses energy. The energy is provided by respiration inside the root hair cells. (You can find out about respiration in Chapter 6.) Energy is needed to produce the shape change in the carrier protein.

Most other cells are also able to carry out active transport. In the human small intestine, for example, glucose can be actively transported from the lumen of the intestine into the cells of the villi.

Practical 2.2 To find the effects of different solutions on plant cells

1. Set up a microscope.
2. Take three clean microscope slides. Label them A, B and C.
3. Put a drop of distilled water onto the centre of slide A.
4. Put a drop of medium concentration sugar solution onto slide B.
5. Put a drop of concentrated sugar solution onto slide C.
6. Peel off a very thin layer of the red epidermis from a rhubarb petiole (leaf stalk). To get good results, it should be as thin as possible (only one cell thick).
7. Cut three squares of this epidermis, each with sides about 5 mm long.
8. Put one square into the drop of solution on each of your three slides.
9. Carefully cover each one with a coverslip. Clean excess liquid from your slides with filter paper.
10. Look at each of your slides under the microscope. Make a labelled drawing of a few cells from each one.

Questions

1. Which part of the cell is coloured red?
2. What has happened to the cells in pure water? Explain your answer.
3. What has happened to the cells in medium concentration sugar solution? Explain your answer.
4. What has happened to the cells in concentrated sugar solution? Explain your answer.

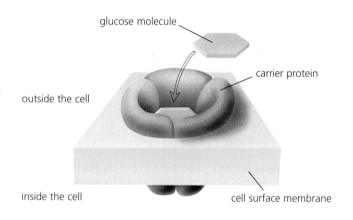

1 The glucose molecule enters the carrier protein.

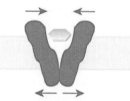

2 The carrier protein changes shape. The energy needed for it to do this is provided by respiration in the cell.

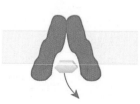

3 The change of shape of the carrier protein pushes the glucose molecule into the cell.

2.8 Active transport.

Practical 2.3 To demonstrate osmosis using eggs

1. Take two fresh hen's eggs. Put each into a beaker and cover them with dilute hydrochloric acid. Leave them for several hours, until all the shell has dissolved.
2. Take the eggs out of the acid, and wash them gently.
3. Label two clean beakers A and B. Put an egg into each.
4. Cover egg A with distilled water. Cover egg B with concentrated salt solution. Leave the eggs in these solutions for a few days.
5. Make a labelled drawing of each egg in its solution.
6. Remove the eggs from their solutions, and examine them. Record any differences between them.

Questions

1. What happened to egg A? Explain your answer.
2. What happened to egg B? Explain your answer.
3. Hen's eggs are not usually immersed in salt solution or water. What substances do you think normally diffuse through the membrane around the egg?

Practical 2.4 The effect of osmosis on potato tissues

Potato tubers are made up of storage tissue made of plant cells. You are going to investigate the effect of osmosis on the tissues of a potato tuber, and also try to estimate their water potential.

1. Collect five containers (such as beakers) and pour into them equal volumes of the five different liquids you have been given. Label each container.
2. Remove the peel from a potato tuber. Using a cork borer, cut as many cylinders from the potato as you can. You will need five sets of identical cylinders, with at least two cylinders in each set – more than that if possible. Each cylinder should be the same length. Choose this length for yourself, depending on the size of your potato – make it as long as possible and certainly no less than 35 mm. Record the initial length of the cylinders.
3. Place each set of potato cylinders into one of the five solutions. Leave them for at least one hour. While you are waiting, read through the rest of these instructions and draw up a results table in which you will record and process your results.
4. Now take the cylinders out of the first of the five solutions. Carefully measure the new length of each cylinder, and record it in your table. Do the same for the cylinders in each of the other five solutions.
5. Calculate the mean (average) length of each set of cylinders after being soaked in their solution. Record this in your table.
6. Calculate the mean change in length for each set of cylinders. Remember to make it clear whether they got longer or shorter.
7. Plot a line graph of mean change in length against concentration of solution. Your y axis will need to show both positive and negative numbers. Ask for help if you are not sure how to do this.
8. Describe and explain the results that you obtained.
   - You should use the term 'water potential' in your explanation.
   - You can use your graph to estimate the water potential of the cells in the potato tissue. If you cannot see how to do this, ask for advice from your teacher.
   - Remember to comment on any anomalous results that you obtained.
9. Although this experiment can give you a good idea of the water potential of the cells in the potato tissue, it is unlikely that the value you obtain will be absolutely correct. Suggest the main limitations of the method and apparatus you have used that could make your results slightly unreliable. (Note – this does not include any mistakes you have made, such as getting your sets of potato cylinders mixed up, or measuring them wrongly!)

3 Enzymes

3.1 Enzymes are biological catalysts.

Many chemical reactions can be speeded up by substances called **catalysts**. A catalyst alters the rate of a chemical reaction, without being changed itself.

Within any living organism, chemical reactions take place all the time. They are sometimes called **metabolic reactions**. Almost every metabolic reaction is controlled by catalysts called **enzymes**.

For example, inside the alimentary canal, large molecules are broken down to smaller ones in the process of digestion. These reactions are speeded up by enzymes. A different enzyme is needed for each kind of food. For example, starch is digested to the sugar maltose by an enzyme called **amylase**. Protein is digested to amino acids by **protease**. These enzymes are also found in plants, for example in germinating seeds, where they digest the food stores for the growing seedling.

Another enzyme which speeds up the breakdown of a substance is **catalase**. Catalase, however, does not work in the alimentary canal. It works *inside* the cells of living organisms – both animals and plants – for example, in liver cells or potato cells. It breaks down hydrogen peroxide to water and oxygen. This is necessary because hydrogen peroxide is produced by many of the chemical reactions which take place inside cells. Hydrogen peroxide is a very dangerous substance, and must be broken down immediately.

Not all enzymes help to break things down. Many enzymes help to make large molecules from small ones. One example of this kind of enzyme is **starch phosphorylase**, which builds starch molecules from glucose molecules inside plant cells.

3.2 Enzymes change substrates to products.

A chemical reaction always involves one substance changing into another. The substance which is present at the beginning of the reaction is called the **substrate**. The substance which is made by the reaction is the **product**. Some examples of substrates and products for reactions catalysed by particular enzymes include:

Substrate	Product	Enzyme
Starch	Maltose	Amylase
Glucose	Starch	Starch phosphorylase
Hydrogen peroxide	Water and oxygen	Catalase

3.3 Enzymes have active sites.

Fig 3.1 shows how an enzyme works. Enzymes are proteins.

Their molecules have a very precise three-dimensional shape. This shape includes a 'dent', which is exactly the right size and shape for a molecule of the enzyme's substrate to fit into. This 'dent' is called the **active site**.

When a substrate molecule slots into the active site, the enzyme 'tweaks' the substrate molecule, pulling it out of shape and making it split into product molecules. The product molecules then leave the active site, which

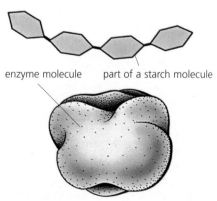

1 Each enzyme has an active site into which its substrate molecule fits exactly. This enzyme is amylase, and its active site is just the right size and shape for a starch molecule.

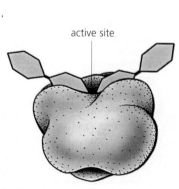

2 The substrate molecule (starch in this case) slots into the active site.

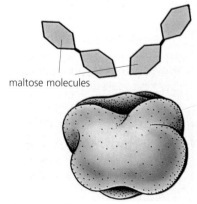

3 The starch molecule is split into maltose molecules. The enzyme is unaltered, and ready to accept another part of the starch molecule.

3.1 How an enzyme works.

is now ready to do the same to another substrate molecule.

This description of how an enzyme works is sometimes called the 'lock and key' hypothesis. The enzyme's active site is the lock, and the substrate is the key. Only the correct key will fit into the lock.

3.4 All enzymes have certain properties.

- **All enzymes are proteins.** They have molecules with a very precise three-dimensional shape, containing an active site.
- **Enzymes are catalysts.** They are unchanged by the reaction they catalyse. Each enzyme molecule can be used over and over again. This means that a small amount of enzyme can catalyse the conversion of a lot of substrate into a lot of product.
- **Enzymes are inactivated by high temperatures.** This is because they are proteins, which are damaged by temperatures above about 40 °C.
- **Enzymes work best at a particular pH.** Most enzymes work best at a pH of about 7. This is also because they are proteins, which are damaged by very acid or very alkaline conditions.
- **Enzymes are specific.** Each enzyme can only convert one kind of substrate molecule into one kind of product molecule. This is because the active site of the enzyme molecule has to be exactly the right shape to allow a substrate molecule to fit into it.

3.5 Temperature affects enzyme-catalysed reactions.

Most chemical reactions happen faster when the temperature is higher. At higher temperatures molecules move around faster, which makes it easier for them to react together. Usually, a rise of 10 °C will double the rate of a chemical reaction (Fig 3.2a).

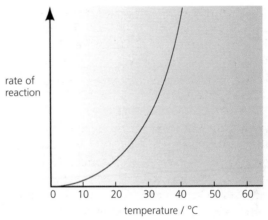

a No enzyme involved.
The rate of reaction doubles with every 10 °C rise in temperature. This is because the molecules which are reacting move faster and have more energy at higher temperatures.

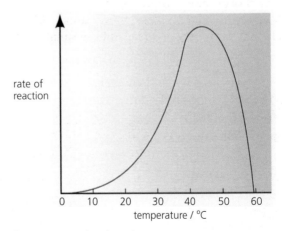

b Enzyme-catalysed reaction.
Between 0 – 40°C, the rate of reaction rises in just the same way as in graph **a** for just the same reasons. But at 40°C the enzyme begins to be damaged, so the reaction slows down. By 60°C, the enzyme is completely destroyed. 40°C is the optimum temperature for this enzyme – the temperature at which the rate of reaction is greatest.

3.2 How temperature affects chemical reactions.

Practical 3.1 The effect of catalase on hydrogen peroxide

Catalase is found in almost any kind of living cell. It catalyses this reaction:

hydrogen peroxide $\xrightarrow{\text{catalase}}$ water + oxygen

1. Read through the instructions. Decide what you will need to observe and measure, and draw a suitable results table.
2. Measure 10 cm³ of hydrogen peroxide into each of five test tubes or boiling tubes.

   TAKE CARE – wear safety goggles.
   (Hydrogen peroxide is a powerful bleach – wash it off with plenty of water if you get it onto your skin.)
3. To each tube, add *one* of the following substances:

   (a) some chopped raw potato
   (b) some chopped boiled potato
   (c) some fruit juice
   (d) a small piece of liver
   (e) some yeast suspension
4. Light a wooden splint, and then blow it out so that it is glowing. Gently push the glowing splint down through the bubbles in your tubes.
5. Record your observations, and explain them as fully as you can.

Most of the chemical reactions happening inside a living organism are controlled or catalysed by enzymes. Enzymes are very sensitive to high temperature. Enzymes from mammals begin to be damaged when the temperature gets to about 40 °C. When this happens to an enzyme, it cannot catalyse its reaction so well, so the reaction slows down. At higher temperatures, the reaction will stop completely because the enzymes are destroyed (Fig 3.2b). Enzymes from plants, such as those involved in seed germination (Section 8.43), usually have lower optimum temperatures than those from mammals, for example 27 °C.

Questions
1. Why do chemical reactions happen faster at high temperatures?
2. What is meant by 'optimum temperature'?
3. Suggest why the optimum temperatures for enzymes found in plants are often lower than those for enzymes found in mammals.

3.6 pH affects enzymes.

Most enzymes only work within a narrow pH range (Fig 3.3). Very acid or very alkaline pHs destroy the shape of the enzyme molecules. However, the enzyme pepsin, which works in the very acid conditions in the stomach, is adapted to function best at a low pH.

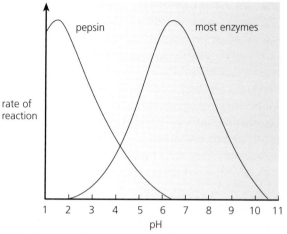

3.3 How pH affects enzyme-catalysed reactions.

Practical 3.2 Investigating the effect of temperature on enzyme activity

Amylase is an enzyme in saliva. It breaks down starch to maltose.

1. Draw a suitable results table.
2. Put 2 cm³ of starch solution into each of four test tubes. Label them A, B, C and D.
3. Put tube A into a refrigerator. Record the temperature of the refrigerator.
4. Put tube B into a test tube rack on your bench. Record the room temperature.
5. Put tube C into a water bath at about 35 °C. Record the temperature of the water bath.
6. Put tube D into a water bath at about 80 °C. Record the temperature of the water bath.
7. Collect about 10 cm³ of amylase solution in a clean boiling tube.
8. Put a drop of iodine solution into each of the cavities on a spotting tile.
9. Add 2 cm³ of amylase solution to each of the four tubes. Record the time at which amylase solution was added to each tube.
10. Stir each tube with its own glass rod.
11. At two-minute intervals, check each tube to see if it still contains starch. Do this by dipping the glass rod into the tube, and then into a spot of iodine solution on the spotting tile. If it turns black, there is still starch there. If it stays brown, the starch has gone.
12. Stop the experiment after 30 minutes.

Questions
1. In which tubes did the starch disappear by the end of the experiment?
2. What had happened to the starch in these tubes?
3. How could you check what the starch had been turned into?
4. Why must each tube have the same amounts of starch, and the same amounts of amylase solution added to them?
5. Why did each tube have its own glass rod?
6. At which temperature did this enzyme seem to work best? Can you explain why?
7. At which temperature did this enzyme seem to work most slowly? Can you explain why?
8. Imagine you are provided with five solutions, each with a different pH. Describe how you could adapt this practical to investigate the effect of pH on the activity of amylase. What results would you expect?

Making use of enzymes

3.7 Enzymes are used in washing powders.

Biological washing powders contain enzymes, as well as detergents. The detergents help greasy dirt to mix with water, so that it can be washed away. The enzymes help to break down other kinds of substances which can stain clothes. They are especially good at removing dirt which contains coloured substances from animals or plants, like blood or egg stains.

Some of the enzymes are **proteases**, which catalyse the breakdown of protein molecules. This helps with the removal of stains caused by proteins, such as blood stains. Blood contains the red protein haemoglobin. The proteases in biological washing powders break the haemoglobin molecules into smaller molecules, which are not coloured, and which dissolve easily in water and can be washed away.

Some of the enzymes are **lipases**, which catalyse the breakdown of fats to fatty acids and glycerol. This is good for removing greasy stains.

The first biological washing powders only worked in warm, rather than hot, water, because the proteases in them had optimum temperatures of about 40 °C. However, proteases have now been developed which can work at much higher temperatures. These proteases have often come from bacteria which naturally live in hot water, in hot springs. This is useful, because the other components of washing powders – which get rid of grease and other kinds of dirt – work best at these higher temperatures.

3.8 Enzymes are often used in the food industry.

Fruit juices are extracted using an enzyme called **pectinase**. Pectin is a substance which helps to stick plant cells together. A fruit such as an apple or orange contains a lot of pectin. If the pectin is broken down, it can be much easier to squeeze juice from the fruit. Pectinase is widely used commercially both in the extraction of juice from fruit, and in making the juice clear rather than cloudy.

Enzymes are sometimes used when making baby foods. Some high-protein foods are treated with **proteases**, to break down the proteins to polypeptides and amino acids. This makes it easier for young babies to absorb the food.

There is a great demand for sugar in the food industry. Not only is it used in making many sweet foods, but it can also be used as a food for microorganisms used in making food substances. (You can read about some of these in Chapter 14.) As well as getting sugar directly from sugar cane and sugar beet, it can be made from starch. This is done by crushing starch-containing materials – perhaps potatoes or grain – with water, and then adding **amylase**. The amylase digests the starch to maltose, making a syrup.

Some sugars are sweeter than others. Fructose, for example – a sugar found in fruits – is sweeter than most other sugars. People who really like sweet things, but are worried about eating too much sugar, may prefer to eat fructose rather than glucose or sucrose, because they can get just as much sweet taste with less sugar. However, most of the sugar we get from plants is either glucose or sucrose. An enzyme called **isomerase** can be used to convert glucose into fructose.

Practical 3.3 Investigating biological washing powders

Design and carry out an experiment to test one or more of the following hypotheses.

(a) A biological washing powder removes egg stains from fabric better than a non-biological washing powder.

(b) Biological washing powders work best at lower temperatures than non-biological washing powders.

(c) Lipases – enzymes which digest fats – help to remove grease stains from fabrics.

If you want to test the protease-containing powders on fabrics, you could first stain the fabrics with a protein-containing stain such as egg.

You could make grease stains using a fat such as lard or butter, mixed up with a little of a red stain called Sudan III.

Your teacher will suggest suitable amounts of enzymes or washing powders to use.

TAKE CARE

Whichever methods you use, do not let the enzymes come into contact with your hands any more than necessary. Remember, you contain a lot of protein and fat!

Question

Name the substrate and product for each of the following enzymes:
pectinase
isomerase
amylase

Practical 3.4 Investigating the use of pectinase in making fruit juice

Design and carry out an experiment to test one or more of the following hypotheses.

(a) You can extract more juice if you add pectinase to the fruit than if you do not.
(b) Juice is extracted more quickly if pectinase is added to the fruit than if it is not.
(c) The effect of pectinase varies on different kinds of fruit, for example apples and pears.
(d) Pectinase has a greater effect on the amount of juice extracted from old fruit than from freshly-picked fruit.
(e) It is more difficult to extract juice, even when using pectinase, from Golden Delicious apples than from other varieties.
(f) People cannot tell the difference between the appearance of juice extracted using pectinase and that of juice extracted without it.
(g) Pectinase added to the extracted juice can make it clear.
(h) Pectinase has an optimum temperature.
(i) Bought fruit juice contains pectinase.

You will find it best to pulp your fruit before trying to extract juice from it. If you are using apples, you can do this by chopping them up roughly, and then blending them with some distilled water in a food processor or other blender. You may then like to heat the blended apple to about 40°C for around 15 minutes, as this speeds up the juice extraction. You can extract the juice by placing the crushed apple in a funnel lined with filter paper, and collecting the juice which drips through.

Your teacher will suggest suitable amounts of pectinase to use.

TAKE CARE

Since you are doing this investigation in a laboratory, and since the pectinase you use may not be food grade, you must *not* taste the fruit juice you make.

Chapter revision questions

1. (a) What is meant by the 'lock and key' hypothesis?
   (b) Use this hypothesis to explain why:
   - each enzyme only works on one type of substrate
   - each enzyme only works well at a particular pH.

2. An experiment was carried out to investigate the effect of the enzyme lipase on fats.
   A solution of lipase was made, and equal volumes of it were added to five test tubes.

   The tubes were treated as follows:
tube 1	kept at 20°C
tube 2	kept at 20°C
tube 3	kept at 0°C
tube 4	kept at 40°C
tube 5	boiled at 100°C

   All five tubes were kept at these temperatures for 15 minutes.
   A small sample of the liquid from each tube was then taken out and added to a small drop of universal indicator solution on a tile. The pH of the liquid in each tube was recorded. Equal volumes of milk (which contains fat) were then added to tubes 2, 3, 4 and 5. Every two minutes, the pH of the contents of the tube were tested as before. The results are shown in the table.

Tube	1	2	3	4	5
Temp. / °C	20	20	0	40	100
Milk added?	no	yes	yes	yes	yes
pH at					
0 minutes	7.0	7.0	7.0	7.0	7.0
2 minutes	7.0	6.8	7.0	6.7	7.0
4 minutes	7.0	6.7	7.0	6.5	7.0
6 minutes	7.0	6.6	7.0	6.3	7.0
8 minutes	7.0	6.6	6.9	6.2	7.0
10 minutes	7.0	6.5	6.9	6.2	7.0

   (a) What is the substrate of the enzyme lipase?
   (b) What is the product when lipase acts on this substrate?
   (c) Explain why, when lipase breaks down its substrate, the pH would be expected to become lower.
   (d) Explain why the pH did not change in tube 1.
   (e) Explain why the pH did not change in tube 5.
   (f) In which tube did lipase act most quickly? Explain why this happened.
   (g) Explain why the results for tubes 2 and 3 were different from each other.
   (h) From these results, what is the optimum temperature for lipase?

4 How animals feed

4.1 All organisms feed.

All living organisms need to take many different substances into their bodies. Some of these may be used to make new parts, or repair old parts. Others may be used to release energy.

Taking in useful substances is called feeding, or **nutrition**. Nutrition can be defined as the obtaining of organic (carbon-containing) substances and mineral ions from which organisms obtain their energy and their raw materials for growth and tissue repair.

4.2 Green plants and animals feed differently.

Green plants take in simple substances, which they get from the air and soil. They use carbon dioxide, water and minerals and build them into more complex materials, such as sugars, in a process called **photosynthesis**. This is described in Chapter 5.

Animals cannot make their own food. They feed on substances which have originally been made by plants. Some animals eat other animals, but all the substances passing from one animal to another were first made by plants.

4.3 Your diet is the food you eat each day.

The food which an animal eats every day is called its **diet**. Most animals need seven kinds of food in their diet. These are:
- carbohydrates
- proteins
- fats
- vitamins
- minerals (inorganic ions)
- water
- roughage.

A diet which contains all of these things, in the correct amounts and proportions, is called a balanced diet (Section 4.17).

Carbohydrates

4.4 Starch and sugars are carbohydrates.

Carbohydrates include starches and sugars (Fig 4.1). Their molecules contain three kinds of atom – carbon (C), hydrogen (H) and oxygen (O). A carbohydrate molecule has about twice as many hydrogen atoms as carbon or oxygen atoms.

4.5 Glucose is a simple sugar.

The simplest kinds of carbohydrate are the simple sugars or **monosaccharides**. **Glucose** is a simple sugar. A glucose molecule is made of six atoms in a ring, with other atoms pointing out from and into the ring (Fig 4.2, overleaf).

The molecule contains six carbon atoms, twelve hydrogen atoms, and six oxygen atoms. To show this, its **molecular formula** can be written $C_6H_{12}O_6$. This formula stands for one molecule of a simple sugar, and tells you which atoms it contains, and how many of each kind.

Although they contain many atoms, simple sugar molecules are very small. They are soluble, and they taste sweet.

4.6 Sucrose is a complex sugar.

If two simple sugar molecules join together, a larger molecule called a complex sugar or **disaccharide** is made (Fig 4.1b). Two examples of complex sugars are **sucrose** (the sugar you add to food or drinks) and **maltose** (malt sugar). Like simple sugars, they are soluble and they taste sweet.

4.7 Starch is a polysaccharide.

If many simple sugars join together, a very large molecule called a polysaccharide is made. Some polysaccharide molecules contain thousands of sugar molecules joined together in a long line. The cellulose of plant cell walls is a polysaccharide and so is starch, which is often found inside plant cells (Fig 4.1c).

Most polysaccharides are insoluble, and they do not taste sweet.

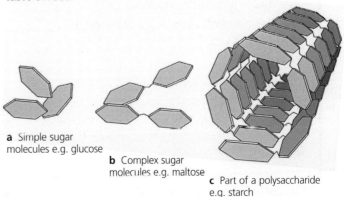

a Simple sugar molecules e.g. glucose

b Complex sugar molecules e.g. maltose

c Part of a polysaccharide e.g. starch

4.1 Carbohydrate molecules.

Practical 4.1 Testing food for carbohydrates

Whenever you are doing any kind of food tests, there are certain procedures you should always follow.

- **Always do a standard test first.** For example, if you are testing foods for simple sugars, begin by testing a known simple sugar, such as glucose. Keep the result of this test, so that you can compare your other results with it.
- **Keep foods completely separate from each other.** This means using clean tubes, spatulas and knives for each kind of food.
- **Use the same volume of reagents for each test.**

To test for simple sugars

1. Draw a results chart.
2. Cut or grind a little of the food into very small pieces. Put these into a test tube. Add some water, and shake it up to try to dissolve it.
3. Add some **Benedict's solution**. Benedict's solution is blue, because it contains copper salts.
4. Boil it. If there is any simple sugar in the food, an orange–red precipitate will form.

This test works because the simple sugar reduces the blue copper salts to a red compound. Sugars which do this are called **reducing sugars**. All simple sugars are reducing sugars, and so are some complex sugars.

To test for starch

There is no need to dissolve the food for this test.

1. Draw a results chart.
2. Put a small piece of the food onto a white tile.
3. Add a drop or two of **iodine solution**. Iodine solution is brown, but it turns bluish black if there is starch in the food.

Questions

1. Why should you cut food into small pieces and try to dissolve it before testing for simple sugars?
2. How could you test a solution to see if it contained iodine?

TAKE CARE

Do not boil tubes directly over a flame – use a beaker as a boiling water bath.

4.8 Animals get energy from carbohydrates.

Carbohydrates are needed for energy (Fig 4.3). One gram of carbohydrate releases 17 kJ (kilojoules) of energy in the body. The energy is released by respiration (Chapter 6).

Questions

1. Which three elements are contained in all carbohydrates?
2. The molecular formula for glucose is $C_6H_{12}O_6$. What does this tell you about a glucose molecule?
3. To which group of carbohydrates do each of these substances belong: (a) glucose, (b) starch, (c) sucrose?
4. Why do animals need carbohydrates?

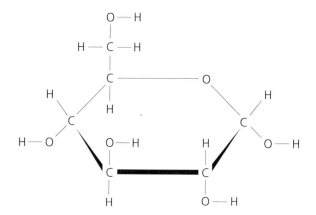

4.2 The structure of a glucose molecule.

CARBOHYDRATES ARE NEEDED FOR ENERGY

4.3 Carbohydrate foods.

Proteins

4.9 Proteins are chains of amino acids.

Protein molecules contain some kinds of atoms which carbohydrates do not. As well as carbon, hydrogen and oxygen, they also contain nitrogen (N) and some have small amounts of sulphur (S).

Like polysaccharides, protein molecules are made of long chains of smaller molecules joined end to end. These smaller molecules are called **amino acids** (Fig 4.4). There are about twenty different kinds of amino acid. Any of these twenty can be joined together in any order to make a protein molecule (Fig 4.5). Each protein is made of molecules with amino acids in a precise order. Even a small difference in the order of amino acids makes a different protein, so there are millions of different proteins which could be made.

Some proteins are soluble, such as haemoglobin, the red pigment in blood. Others are insoluble, such as keratin. Hair and fingernails are made of keratin.

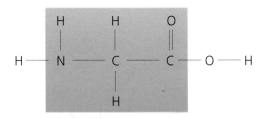

4.4 The structure of an amino acid molecule.

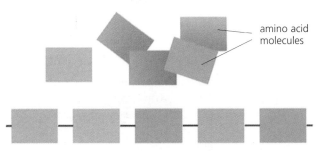

4.5 Protein and amino acid molecules.

Questions

1 Name two elements found in proteins which are not found in carbohydrates.
2 How many different amino acids are there?
3 In what way are protein molecules similar to polysaccharides?
4 Give two examples of proteins.
5 List three reasons why animals need proteins.

4.10 Proteins are used for growth and repair.

Unlike carbohydrates, proteins are not always used to provide energy. Many of the proteins in the food you eat are used for making new cells. New cells are needed for growing, and for repairing damaged parts of the body (Fig 4.6). In particular, **cell membranes** and **cytoplasm** contain a lot of protein.

Proteins are also needed to make **antibodies**. These fight bacteria and viruses inside the body. **Enzymes** are also proteins.

4.6 Protein foods.

Practical 4.2 Testing food for proteins

The biuret test

1 Draw a results chart.
2 Put the food into a test tube, and add a little water.
3 Add some potassium hydroxide solution.
4 Add two drops of copper sulphate solution.
5 Shake the tube gently. If a purple colour appears, then there is protein present.

TAKE CARE

Wear safety glasses. Potassium hydroxide is an alkali. If you get it on your skin, wash it off immediately with plenty of cold water.

Fats

4.11 Fats are made of glycerol and fatty acids.

Like carbohydrates, fats contain only three kinds of atom – carbon, hydrogen and oxygen. A fat molecule is made of four molecules joined together. One of these is glycerol. Attached to the glycerol are three long molecules called fatty acids (Fig 4.7).

Fats are insoluble in water.

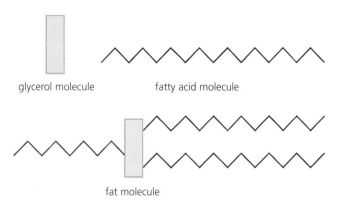

4.7 Fat molecule.

4.12 Animals store energy in fat.

Like carbohydrates and proteins, fats can be used in a cell to release energy. A gram of fat gives about 39 kJ of energy. This is more than twice as much energy as that released by a gram of carbohydrate or protein (Table 4.1).

The extra energy which they contain makes fats very useful for storing energy. Some cells, particularly ones underneath the skin, become filled with large drops of fat or oils. These stores can be used to release energy when needed. This layer of cells is called **adipose tissue**. It also helps to keep heat inside the body – that is, it insulates the body. Animals which live in very cold places, such as seals and some whales, often have especially thick layers of adipose tissue, called blubber.

Practical 4.3 Testing food for fats

The emulsion or ethanol test

Fats will not dissolve in water, but they will dissolve in alcohol. If a solution of fat in alcohol is added to water, the fat forms tiny globules, which float in the water. This is called an emulsion. The globules of fat make the water look milky.

1. Draw a results chart.
2. Chop or grind a small amount of food, and put some into a very clean, dry test tube. Add some absolute (pure) alcohol (ethanol). Shake it thoroughly.
3. Put some distilled water in another tube.
4. Pour some of the liquid part, but not any solid, from the first tube into the water. A milky appearance shows that there is fat in the food.

Questions

1. Which three elements are contained in all fats?
2. List two reasons why animals need fats in their diet.

4.13 Vitamins are needed in small amounts.

Vitamins are organic substances which are only needed in tiny amounts. If you do not have enough of a vitamin, you may get a deficiency disease. Table 4.2 shows two of the most important ones.

Table 4.1 Carbohydrates, proteins and fats.

	Carbohydrates	Proteins	Fats
Elements which they contain	C, H, O	C, H, O, N	C, H, O
Smaller molecules of which they are made	Simple sugars (monosaccharides)	Amino acids	Fatty acids and glycerol
Solubility in water	Sugars are soluble; polysaccharides are not very soluble, often completely insoluble	Some are soluble, some insoluble	Insoluble
Why living organisms need them	Easily available energy (17 kJ/g)	Making cells, antibodies, enzymes; only used for energy when other stores have run out (17 kJ/g)	Storage of energy (39 kJ/g), insulation
Some foods that contain them	Bread, cakes, potatoes, rice, yams	Meat, fish, eggs, milk, cheese, peas, beans	Butter, lard, margarine, oil, fat meat, peanuts

Table 4.2 Vitamins.

Vitamin	Foods that contain it	Why it is needed	Deficiency disease
C	Citrus fruits (e.g. oranges, limes), raw vegetables	Keeps tissues in good repair	Scurvy, which causes pain in joints and muscles, and bleeding from gums and other places; this used to be a common disease of sailors, who had no fresh vegetables for long voyages
D	Butter, egg yolk; can be made by the skin when sunlight shines on it	Helps calcium and phosphate to be used for making bones	Rickets, which causes bones to become soft and deformed

4.14 Minerals are needed in small amounts.

Minerals are inorganic substances. Only small amounts of them are needed in a diet. Table 4.3 shows two of the most important ones.

4.15 Water dissolves substances in cells.

Inside every living organism, chemical reactions are going on all the time. These reactions are called metabolism. Metabolic reactions can only take place if the chemicals which are reacting are dissolved in water. This is one reason why water is so important to living organisms. If their cells dry out, the reactions stop, and the organism dies.

Water is also needed for other reasons. For example, plasma, the liquid part of blood, must contain a lot of water, so that substances like glucose can dissolve in it. Dissolved substances are transported around the body.

4.16 Roughage keeps the alimentary canal moving.

Roughage, or fibre, is food which cannot be digested. It goes right through the digestive system from one end to the other, and is egested in the faeces (Section 4.43).

Roughage helps to keep the alimentary canal working properly. Food moves through the alimentary canal because the muscles contract and relax to squeeze it along. This is called peristalsis (Fig 4.17).

The muscles are stimulated to do this when there is food in the alimentary canal. Soft foods do not stimulate the muscles very much. The muscles work more when there is harder, less digestible food, like roughage, in the alimentary canal. Roughage keeps the digestive system in good working order, and helps to prevent constipation.

All plant food, such as fruit and vegetables, contains roughage. This is because the plant cells have cellulose cell walls. Some plants also contain lignin. Humans cannot digest cellulose or lignin.

One common form of roughage is the outer husk of cereal grains, such as oats, wheat and barley. This is called bran. Some of this husk is also found in wholemeal bread. Brown, or unpolished, rice is also a good source of roughage.

Questions

1 What two different types of food are only needed in small amounts?
2 Why are fresh fruit and vegetables important parts of a healthy diet?

Table 4.3 Minerals.

Mineral element	Foods that contain it	Why it is needed	Deficiency disease
Calcium (Ca)	Milk, cheese, bread	For bones and teeth	Brittle bones and teeth
Iron (Fe)	Liver, egg yolk	For making haemoglobin, the red pigment in blood which carries oxygen	**Anaemia**, which means there are not enough red blood cells, so the tissues are short of oxygen and cannot release energy

> **Questions**
> 1 Which vitamin prevents rickets?
> 2 Why do you need iron in your diet?
> 3 What is metabolism?
> 4 Why do organisms die if they do not have enough water?
> 5 Describe how food is moved along the alimentary canal.
> 6 Plant foods contain a lot of roughage. Explain.

Balanced diet

4.17 Diets should provide the right amount of energy.

Every day, a person uses up energy. The amount you use partly depends on how old you are, which sex you are and what job you do. A few examples are shown in Fig 4.8.

The energy you use each day comes from the food you eat. If you eat too much food, some of the extra will probably be stored as fat. If you eat too little, you may not be able to obtain as much energy as you need. This will make you feel tired.

All food contains some energy. Scientists have worked out how much energy there is in each particular kind of food. You can look up this information in reference books. A few examples are given in Table 4.4.

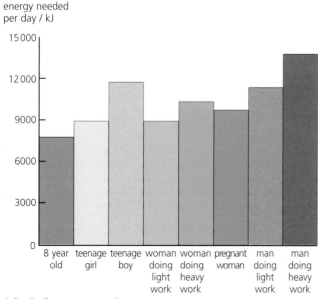

4.8 Daily energy requirements.

4.18 Diets should contain a variety of food.

As well as providing you with energy, food is needed for many other reasons. To make sure that you eat a balanced diet you must eat food containing carbohydrate, fat and protein. You also need each kind of vitamin and mineral, roughage and water. If you miss out any of these things, your body will not be able to work properly.

The staple foods of many countries are carbohydrate foods. Rice, the staple food of many countries, is largely a carbohydrate food. So are potatoes, the staple food for much of Europe. These foods also contain plenty of fibre and some vitamins and minerals. However, they may not contain very much protein.

4.19 Too much saturated fat increases the risk of heart disease.

The kind of fat found in animal foods is called **saturated fat**. These foods also contain **cholesterol**. Some research suggests that people who eat a lot of saturated fat and cholesterol are more likely to get heart disease than people who do not. This is because fat deposits build up on the inside of arteries, making them stiffer and narrower. If this happens in the arteries supplying the heart muscle with blood, then not enough blood can get through.

Table 4.4 Energy content of some different kinds of food.

Food	kJ per 100g	Food	kJ per 100g
Breakfast foods		**Main meals**	
Cornflakes	1567	Stewed steak	932
Oatmeal	1698	Roast chicken	599
Boiled egg	612	Fish (fresh)	340
Brown bread	948	Fish (dried, salt)	857
White bread	991	Fried liver	1016
Milk	272	Sardines	906
Sugar	1680	Cheddar cheese	1682
Marmalade	315	Cottage cheese	402
Unsweetened fruit juice	143	Baked beans	270
		Cabbage	66
		Carrots	98
Desserts		Lettuce	36
Pawpaw	160	Peas	161
Bananas	326	Boiled potatoes	339
Melon	96	French fries	1065
Oranges	150	Tomatoes	60
Canned peaches	373	Rice	1536
Ice cream	698	Spaghetti or noodles	1612
Custard	496	Lentils (dry)	1293
Snacks			
Chocolate	2214		
Fruit yoghurt	405		
Plain biscuits	1925		
Chocolate biscuits	2197		
Roast peanuts	2364		

The heart muscles run short of oxygen and cannot work properly. The deposits can also cause a blood clot to be formed, which results in a heart attack.

Dairy products such as milk, cream, butter and cheese contain a lot of saturated fat. So do red meat and eggs. But vegetable oils are usually **unsaturated fats**. These do not cause heart disease. So it is sensible to use these instead of animal fats when possible.

Vegetable oil can be used for frying instead of butter or lard. Polyunsaturated spreads can be used instead of butter.

Fish and white meat such as chicken do not contain much saturated fat, so eating more of these and less red meat may help to cut down the risk of heart disease.

4.20 Obesity causes health problems.

People who take in more kilojoules than they use up get fat. Being very fat is called **obesity**. Obesity is dangerous to health. Obese people are more likely to get heart disease, strokes, diabetes and many other problems.

Most people can control their weight by eating normal, well-balanced meals and taking regular exercise. Crash diets are not a good idea, except for someone who is very overweight. Although a person may manage to lose a lot of weight quickly, he will almost certainly put it on again once he stops dieting.

4.21 Starvation and malnutrition are different.

In many countries in the world, there is no danger of people suffering from obesity. In some parts of Africa, for example, several years of drought can mean that the harvests do not provide enough food to feed all the people. Despite help from developed countries, many people have died from starvation.

Even if there is enough food to keep people alive, they may suffer from **malnutrition**. Malnutrition is caused by not eating a balanced diet. One common form of malnutrition is **kwashiorkor**. This is caused by a lack of protein in the diet. It is most common in children between the ages of nine months and two years, after they have stopped feeding on breast milk.

Kwashiorkor is often caused by poverty, because the child's mother does not have any high-protein food to give to it. But sometimes it is caused by a lack of knowledge about the right kinds of food that should be eaten.

Children suffering from kwashiorkor are always underweight for their age. But they may often look quite fat, because their diet may contain a lot of carbohydrate. If they are put onto a high-protein diet, they usually begin to grow normally again.

4.22 Food additives can be helpful.

A food additive is something which is added to the food for reasons other than nutrition. The main kinds of food additives, and the reasons for adding them to food, are listed in Table 4.5 (overleaf).

In Europe, each permitted food additive is given an E number. This was meant to reassure people, because it shows that each additive has been tested and passed as safe. However, many people do not like the idea of anything 'unnatural' being added to food, and so E numbers have come to be viewed with suspicion.

In some cases, this suspicion may be justified. The orange food colouring tartrazine, for example, does appear to cause behavioural problems in some children. Food colourings do not improve the food in any way.

However, many food additives are really very good for us. Ascorbic acid, for example, added to many foods to help it to keep well and not go brown, is vitamin C. And without preservatives, there would be many more cases of food poisoning each year.

4.23 There are problems with world food supplies.

It has been calculated that more than enough food is produced on Earth to provide every single person with more than enough for their needs. Yet many people do not get enough food. Each year, many people – both children and adults – die because they have an inadequate diet.

The fundamental problem is that, while some parts of the world produce more than enough food for the people that live there, in other parts of the world nowhere near enough food is produced. Food is distributed unequally on our planet. Although large amounts of food are transported from one area to another, this is still not enough to supply enough food to everybody.

Famines can occur for many different reasons. Often, the main cause is the weather. If an area suffers drought for several years in succession, then it becomes impossible for the people to grow crops. Their animals die, too. Sometimes, however, the problem is exactly the opposite – so much rain falls that it causes flooding, again preventing crops from growing. Sometimes, even though the weather remains normal, the human population may grow so large that the land on which they live can no longer provide enough food for them. Sometimes, wars raging in an area prevent people from working the land and harvesting their crops.

When the world becomes aware that an area is suffering from famine, other countries are usually very willing to donate food supplies to the people.

Hopefully, this will only need to happen for a relatively short time, until things improve and people can plant their crops and become self-sufficient again. Most people would much prefer this, rather than having to rely on hand-outs of food.

> **Questions**
>
> 1 How many kilojoules a day are needed by:
>   (a) a teenage girl,
>   (b) a man doing heavy work?
> 2 Which kinds of food contain saturated fat, and why should they be avoided?
> 3 Explain the difference between starvation and malnutrition.

Digestion

4.24 Digestion makes food easier to absorb.

An animal's alimentary canal is a long tube running from one end of its body to the other (Fig 4.9). Before food can be of any use to the animal, it has to get out of the alimentary canal and into the bloodstream. This is called **absorption**. To be absorbed, molecules of food have to get through the walls of the alimentary canal. They need to be quite small to be able to do this.

The food you eat usually contains some large molecules. Before they can be absorbed, they must be broken down into small ones. This is called **digestion**.

4.25 Not all foods need digesting.

Large carbohydrate molecules, such as polysaccharides, have to be broken down to simple sugars. Proteins are broken down to amino acids. Fats are broken down to fatty acids and glycerol (Fig 4.10).

Simple sugars, water, vitamins and minerals are small molecules, and can be absorbed just as they are.

Table 4.5 Some examples of food additives.

Type of additive	Example	Types of food	Function of additive	Notes
Flavourings	Monosodium glutamate	Soups, stock cubes, many convenience foods	Enhances the flavour of savoury foods	An amino acid which occurs naturally in many foods; has been used in Chinese cooking for centuries
	Vanillin	Desserts, cakes, chocolate	Gives vanilla taste	Natural vanillin comes from the pods of the vanilla orchid; but most is now made artificially
Colourings	Tartrazine	Sweets, drinks	Gives yellow or orange colour	Tartrazine can cause hyperactivity in some children
	Caramel	Sweets, drinks, soups	Gives brown colour	Caramel is made by heating sugar
Preservatives	Sulphur dioxide	Fruit juice, dried fruit	Kills bacteria and preserves the vitamin C content of the food	
	Sodium nitrite	Meat products, e.g. sausages	Stops growth of harmful bacteria	There is some evidence that large amounts of nitrite in the diet may increase the risk of cancer
	Ascorbic acid	Fruits, meat	Stops browning – it is an anti-oxidant	Ascorbic acid is vitamin C; it is also used as a flour improver, as it helps bread dough to rise
Emulsifiers	Lecithin	Powdered milk	Stops oil and water separating out into layers	

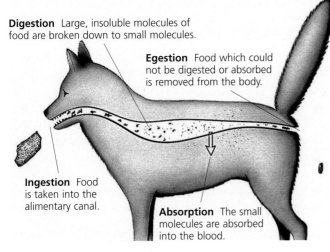

Digestion Large, insoluble molecules of food are broken down to small molecules.

Egestion Food which could not be digested or absorbed is removed from the body.

Ingestion Food is taken into the alimentary canal.

Absorption The small molecules are absorbed into the blood.

4.9 How an animal deals with food.

4.26 Digestion may be mechanical and chemical.

Often the food an animal eats is in quite large pieces. These need to be broken up by teeth, and by the churning movements of the alimentary canal. This is called mechanical digestion.

Once any pieces of food have been ground up, the large molecules present are then broken down into small ones. This is called chemical digestion. It involves a chemical change from one sort of molecule to another. Enzymes are involved in this process.

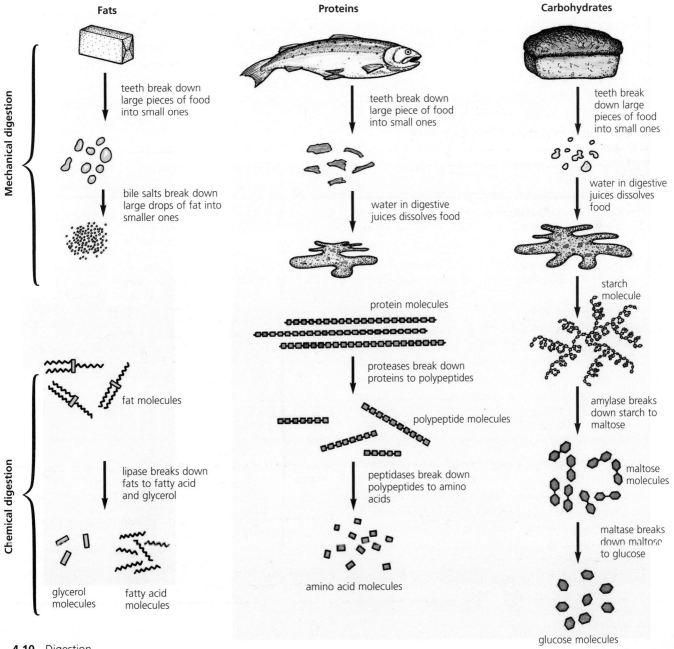

4.10 Digestion.

25

> **Questions**
> 1 What is digestion?
> 2 Name two groups of food which do not need to be digested.
> 3 What does digestion change each of these kinds of food into: (a) polysaccharides, (b) proteins, (c) fats?
> 4 What is meant by chemical digestion?

Digestion in humans – teeth

4.27 Tooth structure is related to function.

Teeth help with the ingestion and mechanical digestion of food. They can be used to bite off pieces of food. They then chop, crush or grind them into smaller pieces. This gives the food a larger surface area, which makes it easier for the enzymes to work. It also helps to dissolve soluble parts of the food. Figs 4.11 and 4.13 show different kinds of human teeth.

The structure of a tooth is shown in Fig 4.12. The part of the tooth which is embedded in the gum is called the **root**. The part which can be seen is the **crown**. The crown is covered with **enamel**.

Enamel is the hardest substance made by animals. It is very difficult to break or chip it. However, it can be dissolved by acids. Bacteria will feed on sweet foods left on the teeth. This makes acids, which dissolve the enamel and decay then sets in.

Under the enamel is a layer of **dentine**, which is rather like bone. This is also quite hard, but not as hard as enamel. It has channels in it, which contain living cytoplasm.

In the middle of the tooth is the **pulp cavity**. It contains nerves and blood vessels. These supply the cytoplasm in the dentine with food and oxygen.

The root of the tooth is covered with **cement**. This has fibres growing out of it. These attach the tooth to the jawbone, but allow it to move slightly when biting or chewing.

4.28 Mammals have different types of teeth.

One of the ways in which mammals differ from other animals is that they have different kinds of teeth. Most mammals have four kinds (Figs 4.11, 4.13). **Incisors** are the sharp-edged, chisel-shaped teeth at the front of the mouth. They are used for biting off pieces of food. **Canines** are the more pointed teeth at either side of the incisors. **Premolars** and **molars** are the large teeth towards the back of the mouth. They are used for chewing food. The molars right at the back are sometimes called wisdom teeth. They do not grow until much later than the others.

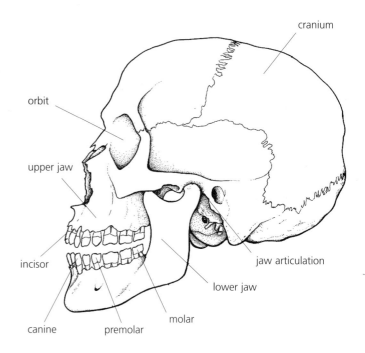

4.11 A human skull.

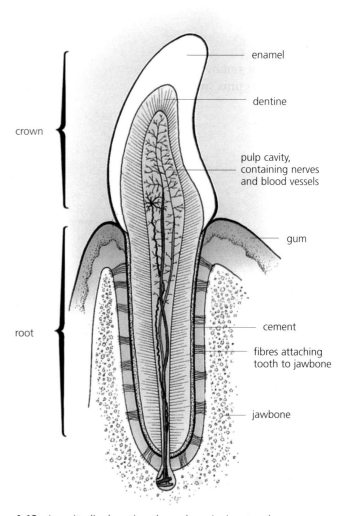

4.12 Longitudinal section through an incisor tooth.

4.29 A mammal has two sets of teeth in its life.

Mammals also differ from other animals in having two sets of teeth. The first set is called the **milk teeth** or **deciduous teeth**. In humans, these start to grow through the gum, one or two at a time, when a child is about five months old. By the age of eighteen or twenty months, most children have a set of 20 teeth.

The milk teeth begin to fall out when the child is about seven years old. They are all replaced by new ones, and twelve new teeth also grow, making up the complete set of permanent teeth. There are 32 altogether. Most people have all their **permanent teeth** by about seventeen years of age.

4.30 Plaque causes tooth decay.

Tooth decay and gum disease are common problems. Both are caused by **bacteria**. You have large numbers of bacteria living in your mouth, most of which are harmless. However, some of these bacteria, together with substances from your saliva, form a sticky film over your teeth, especially next to the gums and in between the teeth. This is called **plaque**.

Plaque is soft and easy to remove at first, but if it is left it hardens to form tartar, which cannot be removed by brushing.

Gum disease

If plaque is not removed, the bacteria in it may infect the gums. The gums swell, become inflamed, and may bleed when you brush your teeth. This is usually painless, but if the bacteria are allowed to spread they may work down around the root of the tooth. The tooth becomes loose, and needs removing (Fig 4.14).

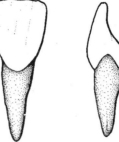
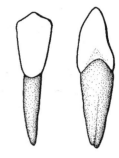

Incisors are chisel shaped, for biting off pieces of food.

Canines are very similar to incisors in humans.

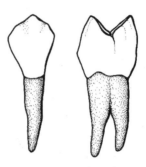

Premolars have wide surfaces, for grinding food.

Molars, like premolars, are used for grinding food.

4.13 Types of human teeth.

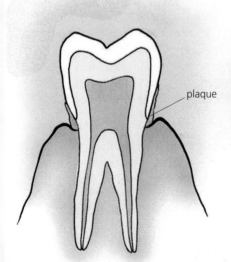
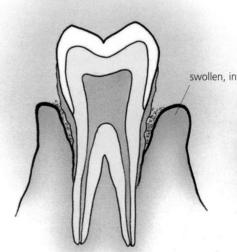
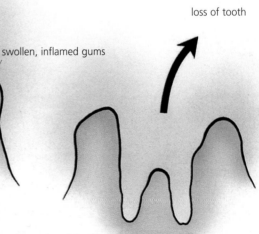

1 Plaque builds up around the edges of teeth and gums.

2 If the plaque is not removed, the bacteria may work down around the roots of the tooth.

3 The tooth is loosened and may fall out or have to be removed.

4.14 Gum disease.

Tooth decay

If sugar is left on the teeth, bacteria in the plaque will feed on it, changing it into **acid**. The acid gradually dissolves the enamel covering the tooth, and works its way into the dentine (Fig 4.15). Dentine is dissolved away more rapidly than the enamel. If nothing is done about it, the tooth will eventually have to be taken out.

4.31 Tooth decay and gum disease can be prevented.

There are several easy things which you can do to keep your teeth and gums healthy and free from pain.

1 Don't eat too much sugar.

If you never eat any sugar, you will not have tooth decay. But nearly everyone enjoys sweet foods, and if you are careful you can still eat them without damaging your teeth. The rule is to eat sweet things only once or twice a day, preferably with your meals. The worst thing you can do it to suck or chew sweet things all day long. And don't forget that many drinks also contain a lot of sugar.

2 Use a fluoride toothpaste regularly.

Fluoride makes your teeth more resistant to decay. Drinking water which contains fluoride, or brushing teeth with a fluoride toothpaste, makes it much less likely that you will have to have teeth filled or extracted.

Regular and thorough brushing also helps to remove plaque, which will prevent gum disease and reduce decay.

3 Make regular visits to a dentist.

Regular dental check ups will make sure that any gum disease or tooth decay is stopped before it really gets a hold.

4.32 Fluoride may be added to drinking water.

In some parts of the world, fluoride is added to drinking water. This is done because fluoride helps to reduce tooth decay.

However, some people do not like this idea. They feel that they ought to be able to make the choice about whether the water they drink contains extra fluoride or not. Also, too much fluoride can cause problems of its own. It may make the teeth go black.

On the other hand, there is no doubt that added fluoride in the water supply does reduce the incidence of tooth decay. This is especially true in areas where fluoride levels in the water are naturally very low.

It is difficult for governments to make a decision that will make everyone happy. Many people do not like the idea of anything unnecessary being added to their water. They say that they can get the same effect by brushing their teeth with a fluoride-containing toothpaste. Others think that it is a good idea to add fluoride, because this means that even people who cannot afford expensive toothpastes, or who do not look after their teeth properly, will benefit.

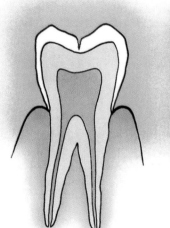
1 Particles of sugary foods get trapped in cracks in the teeth.

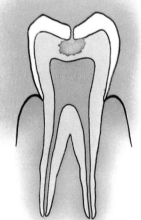
2 Bacteria feeding on the sugar form acids, which dissolve a hole in the enamel and dentine.

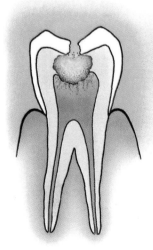
3 There are nerves in the pulp cavity, so the tooth becomes very painful if the infection gets this far.

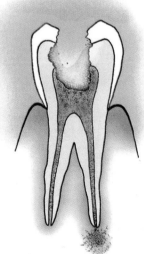
4 The infection can spread rapidly through the pulp cavity, and may form an abscess at the root of the tooth.

4.15 Tooth decay.

> ### Questions
> 1 What are incisors, and what are they used for?
> 2 Describe two ways in which mammals' teeth differ from those of other animals.
> 3 What is plaque?
> 4 Explain how plaque can cause
>   (a) gum disease
>   (b) tooth decay.

Digestion in humans – the alimentary canal

4.33 The alimentary canal is a muscular tube.

The alimentary canal (Fig 4.16) is a long tube which runs from the mouth to the anus. The wall of the tube contains muscles, which contract and relax to make food move along. This movement is called **peristalsis** (Fig 4.17).

Sometimes, it is necessary to keep the food in one part of the alimentary canal for a while, before it is allowed to move into the next part. Special muscles can close the tube completely in certain places. They are called **sphincter muscles**.

To help the food to slide easily through the alimentary canal, it is lubricated with **mucus**. Mucus is made in goblet cells, which occur all along the alimentary canal.

Each part of the alimentary canal has its own part to play in the digestion, absorption and egestion of food.

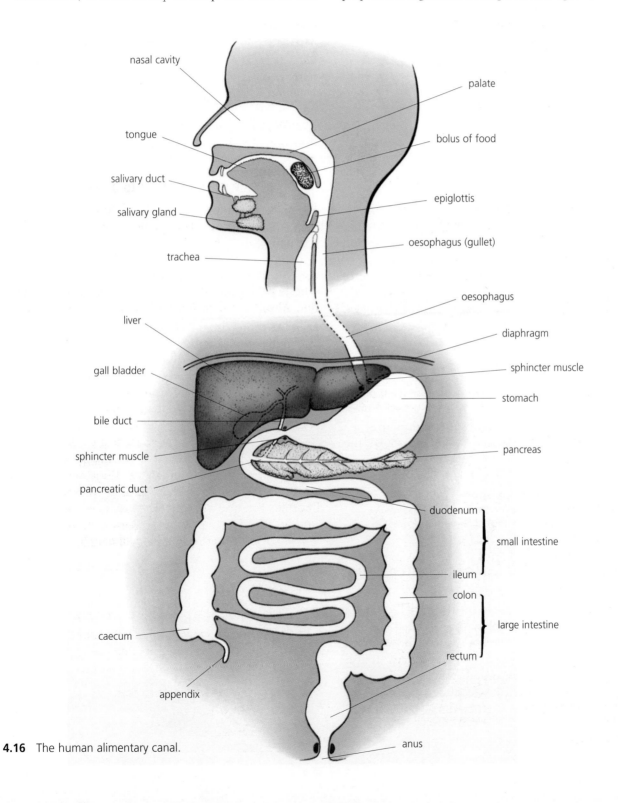

4.16 The human alimentary canal.

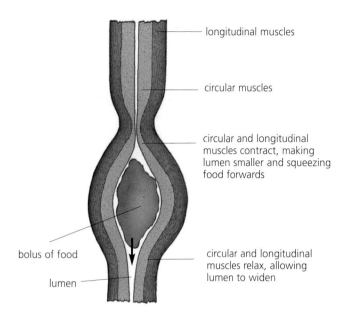

4.17 Peristalsis.

4.34 In the mouth, food is mixed with saliva.

Food is ingested using the teeth, lips and tongue. The teeth then bite or grind the food into smaller pieces. The tongue mixes the food with saliva, and forms it into a **bolus**. The bolus is then swallowed.

Saliva is made in the **salivary glands**. It is a mixture of water, mucus and the enzyme amylase. Amylase begins to digest starch in the food to maltose. Usually, it does not have time to finish this because the food is not kept in the mouth for very long.

4.35 The oesophagus carries food to the stomach.

There are two tubes leading down from the back of the mouth. The one in front is the trachea or windpipe, which takes air down to the lungs. Behind the trachea is the oesophagus, which takes food down to the stomach.

When you swallow, a piece of cartilage covers the entrance to the trachea. It is called the epiglottis, and it stops food from going down into the lungs.

A sphincter muscle at the bottom of the oesophagus opens to let the food into the stomach.

4.36 The stomach stores food and digests proteins.

The stomach has strong, muscular walls. The muscles contract and relax to churn the food and mix it with the enzymes and mucus. The mixture is called **chyme**.

Like all parts of the alimentary canal, the stomach wall contains goblet cells which secrete mucus. It also contains other cells which produce an enzyme called **pepsin**, and others which make **hydrochloric acid**. These are situated in pits in the stomach wall called **gastric pits** (Fig 4.18).

Pepsin is a protease. It begins to digest proteins by breaking them down into polypeptides. Pepsin works best in acid conditions. The acid also helps to kill any bacteria in the food.

The stomach can store food for quite a long time. After one or two hours, the sphincter at the bottom of the stomach opens and lets the chyme into the duodenum.

4.37 The small intestine is very long.

The small intestine is the part of the alimentary canal between the stomach and the colon. It is about 5 m long. It is called the small intestine because it is quite narrow.

Different parts of the small intestine have different names. The part nearest to the stomach is the **duodenum**. The part nearest to the colon is the **ileum**.

4.38 Pancreatic juice flows into the duodenum.

Several enzymes are secreted into the duodenum. They are made in the **pancreas**, which is a cream coloured gland, lying just underneath the stomach. A tube called the pancreatic duct leads from the pancreas into the duodenum. Pancreatic juice, which is a fluid made by the pancreas, flows along this tube.

This fluid contains many enzymes. One is **amylase**, which breaks down starch to maltose. Another is **trypsin**, which is a protease and breaks down proteins and polypeptides to amino acids. Another is **lipase**, which breaks down fats to fatty acids and glycerol.

These enzymes do not work well in acid environments, but the chyme which has come from

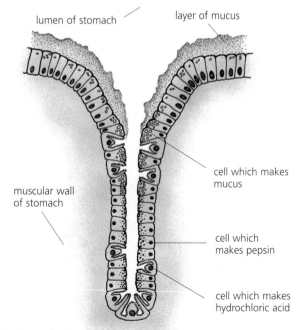

4.18 A gastric pit.

the stomach contains hydrochloric acid. Pancreatic juice contains **sodium hydrogencarbonate**, which neutralises the acid.

4.39 Bile helps to digest fats.

As well as pancreatic juice, another fluid flows into the duodenum. It is called **bile**. Bile is a yellowish green, watery liquid. It is made in the liver, and then stored in the gall bladder. It flows to the duodenum along the bile duct.

Bile does not contain any enzymes. It does, however, help to digest fats. It does this by breaking up the large drops of fat into very small ones, making it easier for the lipase in pancreatic juice to digest them. This is called **emulsification**, and it is done by salts in the bile called bile salts.

Bile also contains yellowish pigments. These are made by the liver when it breaks down old red blood cells. The pigments are made from the haemoglobin. They are not needed by the body, so they are eventually excreted in the faeces.

4.40 Digestion is completed in the small intestine.

As well as receiving enzymes made in the pancreas, the small intestine makes some enzymes itself. They are made by cells in its walls.

The inner wall of the small intestine – the duodenum and ileum – is covered with millions of tiny projections. They are called **villi** (Figs 4.19 and 4.20). Each villus is about 1mm long. It is the cells covering the villi which make the enzymes. The enzymes do not come out into the lumen of the small intestine. They stay attached to the cell surface membranes of the cells lining the small intestine. These enzymes complete the digestion of food.

Maltase breaks down maltose to glucose. **Sucrase** breaks down sucrose to glucose and fructose. **Lactase** breaks down lactose to glucose and galactose. These three enzymes are all carbohydrases. There are also **proteases**, which finish breaking down any polypeptides into amino acids. **Lipase** completes the breakdown of fats to fatty acids and glycerol.

4.41 Digested food is absorbed in the small intestine.

By now, most carbohydrates have been broken down to simple sugars, proteins to amino acids, and fats to fatty acids and glycerol.

These molecules are small enough to pass through the wall of the small intestine and into the blood (Fig 4.21, overleaf). They pass through by diffusion and active transport. This is called absorption. The small intestine is especially adapted to allow absorption to take place very efficiently. Some of its features are listed in Table 4.6.

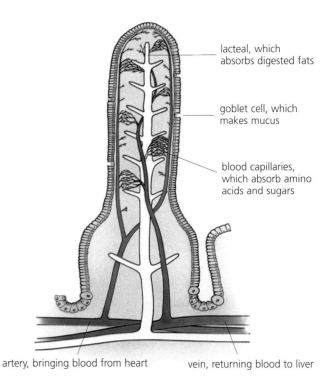

4.19 Longitudinal section through a villus.

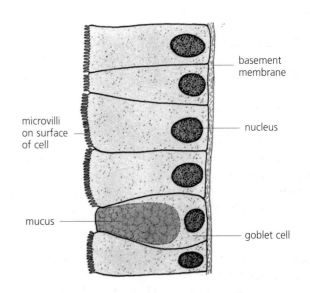

4.20 Detail of villus surface.

4.42 The colon absorbs water.

Not all the food that is eaten can be digested, and this undigested food cannot be absorbed in the small intestine. It travels on, through the caecum, past the appendix and into the colon. In humans, the caecum and appendix have no function. In the colon, water and salt are absorbed.

The colon and rectum are sometimes called the **large intestine**, because they are wider tubes than the duodenum and ileum.

Table 4.6 How the small intestine is adapted for absorbing digested food.

	Feature	How this helps absorption to take place
1	It is very long, about 5 m in an adult.	This gives plenty of time for digestion to be completed, and for digested food to be absorbed as it passes through.
2	It has villi. Each villus is covered with cells which have even smaller projections on them, called microvilli.	This gives the inner surface of the small intestine a very large surface area. The larger the surface area, the faster food can be absorbed.
3	Villi contain blood capillaries.	Digested food passes into the blood, to be taken to the liver and then round the body.
4	Villi contain lacteals, which are part of the lymphatic system.	Fats are absorbed into the lacteals.
5	Villi have walls only one cell thick.	The digested food can easily cross the wall to reach the blood capillaries and lacteals.

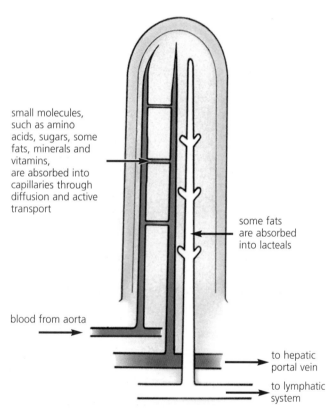

4.21 Diagrammatic section through a villus to show how food is absorbed.

4.43 The rectum temporarily stores undigested food.

By the time the food reaches the rectum, most of the substances which can be absorbed have gone into the blood. All that remains is undigestible food (roughage), bacteria and some dead cells from the inside of the alimentary canal. This mixture forms the faeces, which are passed out at intervals through the anus. The anus has a circular sphincter muscle.

4.44 All absorbed food goes straight to the liver.

After it has been absorbed into the blood, the food is taken to the liver, in the **hepatic portal vein** (Fig 4.22). The liver processes some of it, before it goes any further. Some of the food can be broken down, some converted into other substances, some stored and the remainder left unchanged. The roles of the liver are listed in Table 10.1.

The food, dissolved in the blood plasma, is then taken to other parts of the body where it may become assimilated as parts of the cells.

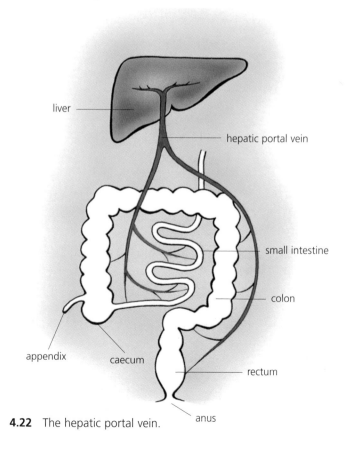

4.22 The hepatic portal vein.

Table 4.7 Summary of chemical digestion in the human alimentary canal.

Part of the canal	Juices secreted	Where made	Enzymes in juice	Substrate	Product	Other substances in juice	Function of other substances in juice
Mouth	Saliva	Salivary glands	Amylase	Starch	Maltose		
Oesophagus	None						
Stomach	Gastric juice	In pits in wall of stomach	Protease (pepsin)	Proteins	Polypeptides	Hydrochloric acid	Acid environment for pepsin; kills bacteria
Duodenum	Pancreatic juice	Pancreas	Amylase	Starch	Maltose	Sodium hydrogen-carbonate ($NaHCO_3$)	Neutralises acidity of chyme, to make an alkaline environment for enzymes
			Protease (trypsin)	Proteins and polypeptides	Amino acids		
			Lipase	Emulsified fats	Fatty acids and glycerol		
	Bile	Liver, stored in gall bladder	None			Bile salts	Emulsify fats
						Bile pigments	Excretory products
Ileum	No juice secreted; enzymes remain in or on the cells covering the villi	By cells covering the villi	Maltase	Maltose	Glucose		
			Sucrase	Sucrose	Glucose and fructose		
			Lactase	Lactose	Glucose and galactose		
			Peptidase	Polypeptides	Amino acids		
			Lipase	Emulsified fats	Fatty acids and glycerol		

All of the digestive juices also contain water. Water is used in splitting the large food molecules into small ones. It also acts as a solvent for the enzymes, substrates and products. The juices also contain mucus, which is a lubricant, and helps to protect the walls of the alimentary canal from being digested by the enzymes. The colon and rectum play no part in digestion. They are concerned with the absorption of water and salts, and with egestion. They are not included in this table.

Chapter revision questions

1 What is a sphincter muscle?
2 Name two places in the alimentary canal where sphincter muscles are found.
3 In which parts of the alimentary canal is mucus secreted?
  Explain why.
4 Name two parts of the alimentary canal where amylase is secreted. What does it do?
5 What is the epiglottis?
6 Why do the walls of the stomach secrete hydrochloric acid?
7 Which two parts of the alimentary canal make up the small intestine?
8 Which two digestive juices are secreted into the duodenum?
9 How do bile salts help in digestion?
10 Name three enzymes made by the cells covering the villi in the small intestine and explain what they do.
11 (a) In which part of the alimentary canal is digested food absorbed?
   (b) Describe three ways in which this part is adapted for absorption.
12 In which part of the alimentary canal is water absorbed?
13 What do faeces contain?
14 A sample of liquid was boiled with Benedict's solution. It stayed blue. Some amylase was then added to another sample of the same liquid, and the mixture was kept at 35°C for ten minutes. It was then boiled with Benedict's solution. This time, an orange–red precipitate was obtained. What did the liquid contain? Explain your answer fully.
15 In some parts of the world, people do not have the enzyme lactase in their alimentary canal. Drinking milk, which contains lactose, can make them ill.
   (a) What does lactase do?
   (b) Suggest how milk could be treated to make it lactose-free.

5 How green plants feed

5.1 Green plants feed autotrophically.

Green plants make their food by using carbon dioxide, water and a variety of minerals to make all the carbohydrates, fats, proteins and vitamins they need.

To make carbohydrates, plants combine carbon dioxide and water. The carbohydrate which is made is **glucose**. At the same time, oxygen molecules are produced.

$$\text{carbon dioxide} + \text{water} \longrightarrow \text{glucose} + \text{oxygen}$$

However, if you mixed together carbon dioxide and water, they would not combine to make glucose and oxygen. They have to be given energy to make them combine. The energy which green plants use for this is **sunlight** energy. The reaction is therefore called **photosynthesis** ('photo' means light, and 'synthesis' means manufacture).

5.2 Chlorophyll absorbs sunlight.

However, sunlight shining onto water and carbon dioxide still will not make them react together in this way. The sunlight energy has to be trapped, and then used in the reaction. Green plants have a substance which does this. It is called **chlorophyll**.

Chlorophyll is the pigment which makes plants look green. It is kept inside the chloroplasts of plant cells. When sunlight falls on a chlorophyll molecule, the energy is absorbed. The chlorophyll molecule then releases the energy. The energy makes carbon dioxide combine with water, with the help of enzymes inside the chloroplast. The glucose that is made contains energy that was originally in the sunlight.

5.3 Photosynthesis is a chemical process.

The full equation for photosynthesis is written like this:

$$\text{carbon dioxide} + \text{water} \xrightarrow[\text{chlorophyll}]{\text{sunlight}} \text{glucose} + \text{oxygen}$$

To show the number of molecules involved in the reaction, a balanced equation needs to be written. Carbon dioxide contains two atoms of oxygen, and one of carbon, so its molecular formula is CO_2. Water has the formula H_2O. Glucose has the formula $C_6H_{12}O_6$. Oxygen molecules contain two atoms of oxygen, and so they are written O_2.

The balanced equation for photosynthesis is this:

$$6CO_2 + 6H_2O \xrightarrow[\text{chlorophyll}]{\text{sunlight}} C_6H_{12}O_6 + 6O_2$$

Questions
1. What kind of energy does a plant use to make carbon dioxide combine with water?
2. What is chlorophyll?
3. What happens when sunlight falls on a chlorophyll molecule?
4. What does a balanced equation show?

Leaves

5.4 Plant leaves are food factories.

Photosynthesis happens inside chloroplasts. This is where the enzymes and chlorophyll are which catalyse and supply energy to the reaction. In a typical plant, most chloroplasts are in the cells in the leaves. A leaf is a factory for making carbohydrates. Leaves are therefore specially adapted to allow photosynthesis to take place quickly and efficiently (Table 5.1).

5.5 Leaf structure is related to function.

A leaf consists of a broad, flat part called the **lamina** (Fig 5.1), which is joined to the rest of the plant by a leaf stalk or **petiole**. Running through the petiole are vascular bundles (Section 7.25), which then form the **veins** in the leaf. These contain tubes, which carry substances to and from the leaf.

Although a leaf looks thin, it is in fact made up of several layers of cells. You can see these if you look at a transverse section (TS) of a leaf under a microscope (Figs 5.2 and 5.3).

The top and bottom of the leaf are covered with a layer of closely fitting cells called the **epidermis** (Figs 5.4 and 5.5). These cells do not contain chloroplasts. Their function is to protect the inner layers of cells in the leaf. The cells of the upper epidermis often secrete a waxy substance, which lies on top of them. It is called the **cuticle**, and it helps to stop water evaporating from the leaf. There is sometimes a cuticle on the underside of a leaf as well.

In the lower epidermis, there are small holes called **stomata** (singular: **stoma**). Each stoma is surrounded by a pair of bean-shaped **guard cells** (Fig 5.4), which can open or close the hole. Guard cells, unlike the other cells in the epidermis, do contain chloroplasts.

The middle layers of the leaf are called the **mesophyll** ('meso' means middle, and 'phyll' means leaf). These cells all contain chloroplasts. The cells nearer to the top of the leaf are arranged like a fence or palisade, and they form the **palisade layer**. The cells beneath them are rounder, and arranged quite loosely, with large air spaces between them. They form the **spongy layer**.

Running through the mesophyll are veins. Each vein contains large, thick-walled **xylem vessels** (Section 7.23) for carrying water, and smaller, thin-walled **phloem tubes** (Section 7.24) for carrying away food which the leaf has made.

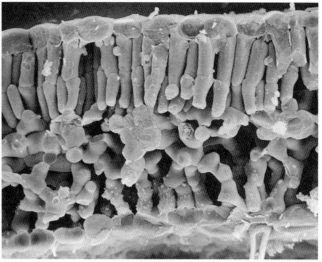

5.3 A scanning electron micrograph showing the cells inside a leaf; notice many air spaces between the cells.

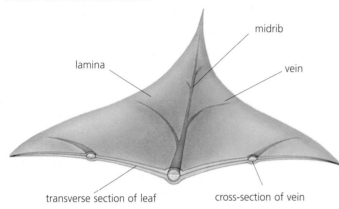

5.1 The structure of a leaf.

Questions
1 What is another name for a leaf stalk?
2 Which kind of cells make the cuticle on a leaf?
3 What is the function of the cuticle?
4 What are stomata?
5 What are guard cells?
6 List three kinds of cells in a leaf which contain chloroplasts, and one kind which does not.

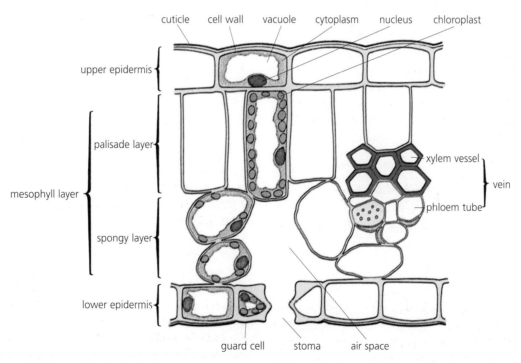

5.2 Transverse section through part of a leaf.

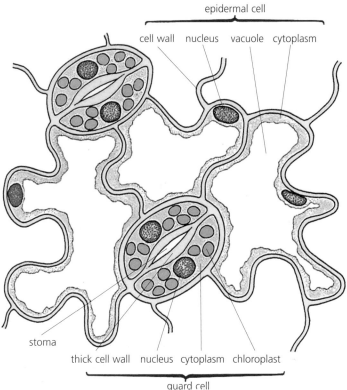

5.4 Surface view of the lower epidermis of a leaf.

5.6 Leaves are adapted to obtain carbon dioxide, water and sunlight.

Carbon dioxide

Carbon dioxide is obtained from the air. There is not very much available, because only about 0.04% of the air is carbon dioxide. Therefore the leaf must be very efficient at absorbing it. The leaf is held out into the air by the stem and the leaf stalk, and its large surface area helps to expose it to as much air as possible.

The cells which need the carbon dioxide are the mesophyll cells, inside the leaf. The carbon dioxide can get into the leaf through the stomata.

It does this by diffusion, which is described in Chapter 2. Behind each stoma is an air space (Fig 5.2) which connects up with other air spaces between the spongy mesophyll cells. The carbon dioxide can therefore diffuse to all the cells in the leaf. It can then diffuse through the cell wall and cell membrane of each cell, and into the chloroplasts.

Water

Water is obtained from the soil. It is absorbed by the root hairs (Section 7.27), and carried up to the leaf in the xylem vessels. It then travels from the xylem vessels to the mesophyll cells by osmosis, which is described in Chapter 2. The path it takes is shown in Figs 5.6 and 5.7.

Sunlight

The position of a leaf and its broad, flat, surface help it to obtain as much sunlight as possible. If you look up through the branches of a tree, you will see that the leaves are arranged so that they do not cut off light from one another more than necessary. Plants which live in shady places often have particularly big leaves.

The cells which need the sunlight are the mesophyll cells. The thinness of the leaf allows the sunlight to penetrate right through it, and reach all the cells. To help this the epidermal cells are transparent, with no chloroplasts.

In the mesophyll cells, the chloroplasts are arranged to get as much sunlight as possible, particularly those in the palisade cells. They can lie broadside on to do this, but in strong sunlight, they often arrange themselves end on. This reduces the amount of light absorbed. Inside them, the chlorophyll is arranged on flat membranes, which exposes as much as possible to the sunlight.

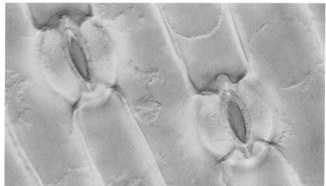

5.5 The lower surface of a leaf, showing the closely-fitting cells of the epidermis; the oval holes are stomata.

5.7 Glucose is used in different ways.

One of the first carbohydrates to be made in photosynthesis is **glucose**. There are several things which may then happen to it.

Energy may be released from glucose in the leaf.

All cells need energy, which they obtain by the process of respiration (Section 6.1). Some of the glucose which a leaf makes will be broken down by respiration inside the leaf cells, to release energy.

Glucose may be turned into starch and stored in the leaf.

Glucose is a simple sugar. It is soluble and is not, therefore, a very good storage molecule. It would dissolve in the water in and around the plant cells, and might be lost from the cell. When dissolved, it would increase the concentration of the solution in the cell, which could damage the cell.

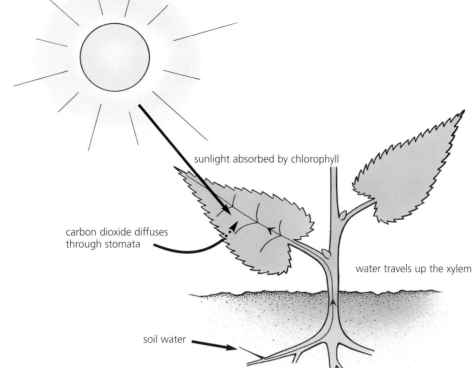

5.6 How the materials for photosynthesis get into a leaf.

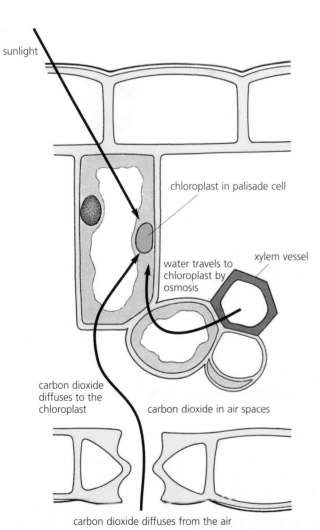

5.7 How the raw materials for photosynthesis get to a palisade cell.

The glucose is therefore converted into **starch** to be stored. Starch is a polysaccharide, made of many glucose molecules joined together. Being such a large molecule, it is not very soluble. It can be made into granules, which can be easily stored inside the chloroplasts.

Glucose may be used to make other organic substances.

The plant can use glucose as a starting point for making all the other organic substances it needs. These include the carbohydrates sucrose and cellulose. Plants also make **oils** (liquid fats).

Plants can also use the sugars they have made in photosynthesis to make **proteins**. To do this, they need **nitrogen**. Unfortunately, even though the air around us is 78% nitrogen, this is completely useless to plants because it is very unreactive. Plants have to be supplied with nitrogen in a more reactive form, usually as **nitrate ions**. They absorb nitrate ions from the soil, through their root hairs, by diffusion and active transport. The nitrate ions combine with glucose to make amino acids. The amino acids are then strung together to form protein molecules.

Another substance that plants make is **chlorophyll**. Once again, they need nitrogen to do this, and also another element – **magnesium**. The magnesium, like the nitrate ions, is obtained from the soil.

Table 5.2 shows what happens to a plant if it does not have enough of these ions. Farmers often add extra mineral ions to the soil in which their crops are growing, to make sure that they do not run short of these essential substances. You can read more about this in Section 16.14.

Sugars may be transported to other parts of the plant.

A molecule has to be small and soluble to be transported easily. Glucose has both of these properties, but it is also rather reactive. It is therefore converted to the complex sugar sucrose to be transported to other parts of the plant. Sucrose molecules are also quite small and soluble, but less reactive than glucose. They dissolve in the sap in the phloem tubes, and can be distributed to whichever parts of the plant need them (Fig 5.8).

The sucrose may later be turned back into glucose again, to be broken down to release energy, or turned into starch and stored, or used to make other substances which are needed for growth.

Questions
1 What are the raw materials needed for photosynthesis?
2 What percentage of the air is carbon dioxide?
3 How does carbon dioxide get into a leaf?
4 How does a leaf obtain its water?
5 Give two reasons why leaves need to have a large surface area.
6 Why are leaves thin?

Practical 5.1 Looking at the epidermis of a leaf

Using a piece of epidermis
1 Using forceps, carefully peel a small piece of epidermis from the underside of a leaf.
2 Put the piece of epidermis into a drop of water on a microscope slide.
3 Spread it out carefully, trying not to let any part of it fold over. Cover it with a coverslip.
4 Look at your slide under the microscope, and make a labelled drawing of a few cells.

Making a nail varnish impression
1 Paint the underside of a leaf with transparent nail varnish. Leave to dry thoroughly.
2 Peel off part of the nail varnish, and mount it in a drop of water on a microscope slide.
3 Spread it out carefully, and cover with a coverslip.
4 Look at your slide under the microscope, and make a labelled drawing of the impressions made by a few cells.
5 Repeat with the upper surface of a leaf.

Questions
1 On which surface of the leaf did you find most stomata?
2 Which of these two techniques for examining the epidermis of a leaf do you consider (a) is easiest, and (b) gives you the best results?
3 There are two kinds of cell in the lower epidermis of a leaf. What are they, and what are their functions?

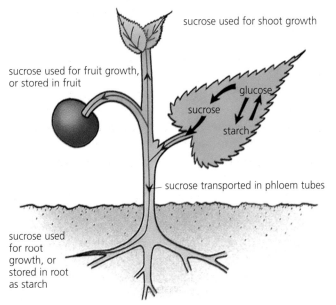

5.8 The products of photosynthesis.

Questions
1 Why is glucose not very good for storing in a leaf?
2 What substances does a plant need to be able to convert glucose into proteins?
3 Why do plants need magnesium?
4 How do parts of the plant such as the roots, which cannot photosynthesise, obtain food?

Table 5.1 Adaptations of leaves for photosynthesis.

Adaptation	Function
Supported by stem and petiole	To expose as much of it as possible to sunlight and air
Large surface area	To expose as much of it as possible to sunlight and air
Thin	To allow sunlight to penetrate to all cells; to allow CO_2 to diffuse in and O_2 to diffuse out as quickly as possible
Stomata in lower epidermis	To allow CO_2 and O_2 to diffuse in and out
Air spaces in spongy mesophyll	To allow CO_2 and O_2 to diffuse to and from all cells
No chloroplasts in epidermis	To allow sunlight to penetrate to mesophyll layer
Chloroplasts containing chlorophyll present in mesophyll layer	To absorb sunlight, to provide energy to combine CO_2 and H_2O
Palisade cells arranged end on	To keep as few cell walls as possible between sunlight and chloroplasts
Chloroplasts in palisade cells arranged broadside on, especially in dim light	To expose as much chlorophyll as possible to sunlight
Chlorophyll arranged on flat membranes inside chloroplast	To expose as much chlorophyll as possible to sunlight
Xylem vessels within short distance of every mesophyll cell	To supply water to chloroplasts for photosynthesis
Phloem tubes within short distance of every mesophyll cell	To take away organic products of photosynthesis

Table 5.2 Mineral salts required by plants.

	Nitrogen	Magnesium
Mineral salt	Nitrates or organic compounds from nitrogen-fixing bacteria	Magnesium salts
Why needed	To make proteins	To make chlorophyll
Deficiency	Poor growth, yellow leaves	Yellowing between veins of leaves

Photosynthesis experiments

5.8 Experiments need controls.

If you do Practicals 5.3, 5.4 and 5.5, you can find out for yourself which substances a plant needs for photosynthesis. In each experiment, the plant is given everything it needs, except for one substance. Another plant is used at the same time. This is a **control**. The control plant is given everything it needs, including the substance being tested for.

Both plants are then treated in exactly the same way. Any differences between them at the end of the experiment, therefore, must be because of the substance being tested.

At the end of the experiment, test a leaf from your experimental plant and your control to see if they have made starch. By comparing them, you can find out which substances are necessary for photosynthesis.

5.9 Plants for photosynthesis experiments must be destarched.

It is very important that the leaves you are testing should not have any starch in them at the beginning of the experiment. If they did, and you found that the leaves contained starch at the end of the experiment, you could not be sure that they had been photosynthesising. The starch might have been made before the experiment began.

So, before doing any of these experiments, you must destarch the plants.

The easiest way to do this is to leave them in a dark cupboard for at least 24 hours. The plants cannot photosynthesise while they are in the cupboard because there is no light. Therefore they use up their stores of starch. To be certain that they are thoroughly destarched, test a leaf for starch before you begin your experiment.

5.10 Iodine solution can stain starch in leaves.

Iodine solution is used to test for starch. A bluish black colour shows that starch is present. However, if you put iodine solution onto a leaf which contains starch, it will not immediately turn black. This is because the starch is right inside the cells, inside the chloroplasts. The iodine solution cannot get through the cell membranes to reach the starch and react with it. Another difficulty is that the green colour of the leaf and the brown iodine solution can look black together.

Therefore before testing a leaf for starch, you must break down the cell membranes, and get rid of the green colour (chlorophyll). The way this is done is described in Practical 5.2. The cell membranes are first broken down by boiling water, and then the chlorophyll is removed by dissolving it out with alcohol.

5.11 Many factors affect rate of photosynthesis.

If a plant is given plenty of sunlight, carbon dioxide and water, the limit on the rate at which it can photosynthesise is its own ability to absorb these materials, and make them react. However, quite often plants do not have unlimited supplies of these materials, and so their rate of photosynthesis is not as high as it might be.

Sunlight

In the dark, a plant cannot photosynthesise at all. In dim light, it can photosynthesise slowly. As light intensity increases, the rate of photosynthesis will increase, until the plant is photosynthesising as fast as it can. At this point, even if the light becomes brighter, the plant cannot photosynthesise any faster (Fig 5.11a).

> **Practical 5.2 Testing a leaf for starch**
>
> 1 Take a leaf from a healthy plant, and drop it into boiling water in a water bath. Leave it for about 30 seconds. (The length of time needed varies for different types of leaves – 30s is about right for *Pelargonium* leaves.)
> 2 Remove the leaf, which will be very soft, and drop it into a tube of alcohol in the water bath (Fig 5.9). Leave it until all the chlorophyll has been dissolved out of the leaf.
> 3 The leaf will now be brittle. Remove it from the alcohol, and dip it into water again to soften it.
> 4 Spread out the leaf on a white tile, and cover it with iodine solution. A black colour shows that the leaf contains starch.
>
>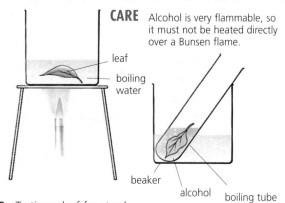
>
> **CARE** Alcohol is very flammable, so it must not be heated directly over a Bunsen flame.
>
> 5.9 Testing a leaf for starch.
>
> ### Questions
> 1 Why was the leaf put into boiling water?
> 2 Why did the alcohol become green?
> 3 Why was the leaf put into alcohol *after* being put into boiling water?

Over the first part of the curve, in Fig 5.11a, between A and B, light is a **limiting factor**. The plant is limited in how fast it can photosynthesise because it does not have enough light. You can show this because when the plant is given more light it photosynthesises faster. Between B and C, however, light is not a limiting factor. You can show this because, even if more light is shone on the plant, it still cannot photosynthesise any faster. It already has as much light as it can use.

> **Practical 5.3 To find out if light is necessary for photosynthesis**
>
> 1 Take a healthy *Pelargonium* plant, growing in a pot. Leave it in a cupboard for a few days, to destarch it.
> 2 Test one of its leaves for starch, to check that it does not contain any.
> 3 Using a folded piece of black paper or aluminium foil, a little larger than a leaf, cut out a shape (Fig 5.10). Fasten the paper or foil firmly over both sides of a leaf on your plant, making sure that the edges are held firmly together. Don't take the leaf off the plant!
> 4 Leave the plant near a warm, sunny window for a few days.
> 5 Remove the cover from your leaf, and test it for starch.
> 6 Make a labelled drawing of the appearance of your leaf after testing for starch.
>
>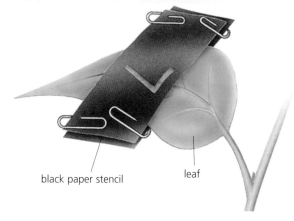
>
> 5.10 To see if light is necessary for photosynthesis.
>
> ### Questions
> 1 Why was the plant destarched before the beginning of the experiment?
> 2 Why was part of the leaf left uncovered?
> 3 What do your results tell you about light and photosynthesis?

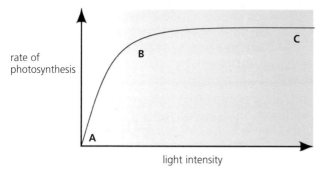

a Light intensity and rate of photosynthesis.

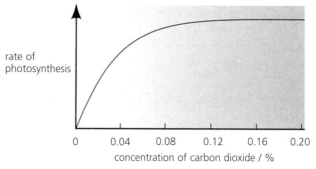

b Carbon dioxide concentration and rate of photosynthesis at high light intensity.

5.11 Some factors affecting the rate of photosynthesis.

Carbon dioxide

Carbon dioxide can also be a limiting factor (Fig 5.11b). The more carbon dioxide a plant is given, the faster it can photosynthesise up to a point, but then a maximum is reached.

> ### Practical 5.4 To see if carbon dioxide is necessary for photosynthesis
>
> 1 Destarch a plant.
> 2 Set up your apparatus as shown in Fig 5.12. Take special care that no air can get into the flasks. Leave the plant in a warm sunny window for a few days.
> 3 Test each treated leaf for starch.
>
>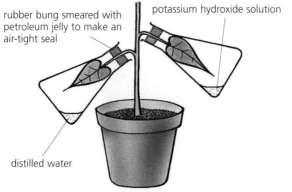
>
> **5.12** To see if carbon dioxide is necessary for photosynthesis.
>
> ### Questions
> 1 Why was potassium hydroxide put in with one leaf, and water with the other?
> 2 Which was the control?
> 3 Why was petroleum jelly put around the tops of the flasks?
> 4 What do your results suggest about carbon dioxide and photosynthesis?

Temperature

The chemical reactions of photosynthesis will only take place very slowly at low temperatures, so a plant can photosynthesise faster on a warm day than a cold one.

Stomata

The carbon dioxide which a plant uses passes into the leaf through the stomata. If the stomata are closed, then photosynthesis cannot take place. Stomata often close if the weather is very hot and sunny, to prevent too much water being lost. On a really hot day, therefore, photosynthesis may slow down or stop for a time.

When plants are growing outside, we cannot do much about changing the conditions that they need for photosynthesis. If a field of sorghum does not get enough sunshine, or is short of carbon dioxide, then it just has to stay that way. But if crops are grown in glasshouses, then it is possible to control the conditions so that they are photosynthesising as fast as possible.

For example, in parts of the world where it is often too cold for good growth of some crop plants, they can be grown in heated glasshouses. This is done, for

> ### Practical 5.5 To see if chlorophyll is necessary for photosynthesis
>
> 1 Destarch a plant with variegated (green and white) leaves.
> 2 Leave your plant in a warm, sunny spot for a few days.
> 3 Test one of the leaves for starch.
> 4 Make a drawing of your leaf before and after testing.
>
> ### Questions
> 1 What was the control in this experiment?
> 2 What do your results tell you about chlorophyll and photosynthesis?

example, with tomatoes. The temperature in the glasshouse can be kept at the optimum level to encourage the tomatoes to grow fast and strongly, and to produce a large yield of fruit that ripens quickly.

Light can also be controlled. In cloudy or dark conditions, extra lighting can be provided, so that light is not limiting the rate of photosynthesis. The kind of lights that are used can be chosen carefully so that they provide just the right wavelengths that the plants need.

Carbon dioxide concentration can also be controlled. Carbon dioxide is often a limiting factor for photosynthesis, because its natural concentration in the air is so very low. In a closed glasshouse, it is possible to provide extra carbon dioxide for the plants.

Practical 5.6 To show that oxygen is produced during photosynthesis

1. Set up the apparatus as shown in Fig 5.13. Make sure that the test tube is completely full of water.
2. Leave the apparatus near a warm, sunny window for a few days.
3. Carefully remove the test tube from the top of the funnel, allowing the water to run out, but not allowing the gas to escape.
4. Light a wooden splint, and then blow it out so that it is just glowing. Carefully put it into the gas in the test tube.

   If it bursts into flame, then the gas is oxygen.

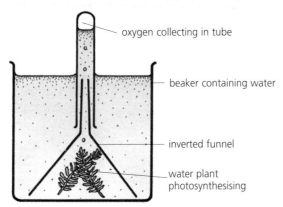

5.13 To show that oxygen is produced during photosynthesis.

Questions

1. Why was this experiment done under water?
2. This experiment has no control. Try to design one.

Practical 5.7 Investigating the effect of light intensity, carbon dioxide concentration or temperature on the rate of photosynthesis

You are going to plan and carry out an experiment to find out if changing any of the three factors listed above makes a plant photosynthesise more quickly or slowly. The following points may help you to design your experiment, and to record and interpret your results.

- You could use apparatus like that in Fig. 5.13. It might be easier to use a smaller container than a beaker, such as a boiling tube. You might get better results if you counted the number of bubbles given off in one minute.
- Make sure that you only try to investigate one of the three factors at once. All other factors should be kept constant. If different people investigate different factors, then you can all share your results later on.
- You could vary the light intensity by placing a lamp at different distances from the plant. Remember, though, that the lamp is giving off heat as well as light. Consider the fact that the plant may need a little time to settle down at each new light intensity.
- You could vary carbon dioxide concentration by adding a little sodium hydrogencarbonate to the water around the plant, which provides it with extra carbon dioxide.
- If at all possible, you should try to take more than one set of readings at each value of light intensity, temperature or carbon dioxide, and then calculate an average.
- You should record your results in a table, and then use them to plot a graph. Think about what will go on the x axis and the y axis.

Chapter revision questions

1. Copy and complete this table.

	obtained from	used for
nitrates	soil, through root hairs	making proteins
water		
magnesium		
carbon dioxide		

Chapter revision questions (continued)

2. Explain the following.
   (a) There is an air space behind each stoma.
   (b) The epidermal cells of a leaf do not have chloroplasts.
   (c) Leaves have a large surface area.
   (d) The veins in a leaf branch repeatedly.
   (e) A leaf containing starch does not turn black immediately when you put iodine solution onto it.
   (f) Chloroplasts have many membranes in them.

3. Which carbohydrate does a plant use for each of these purposes? Explain why.
   (a) transport
   (b) storage

4. Describe how a carbon atom in a carbon dioxide molecule in the air could become part of a starch molecule in a carrot root.

   Mention all the structures it would pass through, and what would happen to it at each stage.

5. Read the following passage carefully, then answer the questions, using both the information in the passage and your own knowledge.

   > White light is made up of all the colours of the rainbow. Sea water acts as a light filter which screens off some of the light energy, starting at the red end of the spectrum. As sunlight travels downwards through the water, first the red light is lost, then green and yellow and finally blue. In very clear water, the blue light can penetrate to a maximum of 1000m. Below this, all is dark.
   >
   > The upper layers of the sea contain a large community of microscopic floating organisms called plankton, many of which are tiny plants known as phytoplankton. These act as a gigantic solar cell, which feeds all the animals of the sea and supplies both them and the atmosphere above with oxygen.
   >
   > Nearer the shore, larger plants are found. Seaweeds grow on rocky shores, brown and green ones high on the shore, and red ones lower down, where they are covered with deep water when the tide is in. The colours of the seaweeds are due to their light-absorbing pigments, not all of which are chlorophyll.

   (a) Why are no green plants found below the upper few hundred metres of the sea?
   (b) Some living organisms are found in the permanently dark depths of the oceans. What might they feed on?
   (c) What are phytoplankton?
   (d) Explain as fully as you can the last sentence of paragraph 2, 'These act as ... with oxygen.'
   (e) Chlorophyll is a green pigment. Which colours of light does it (i) absorb, and (ii) reflect?
   (f) What colour light would you expect the pigment of red seaweeds to absorb?
   (g) Why are red seaweeds normally found lower down the shore than green ones?

6. An experiment was performed to find out how fast a plant photosynthesised as the concentration of CO_2 in the air around it was varied. The results were as follows.

CO_2 concentration / % by volume in air	Rate of photosynthesis in arbitrary units	
	low light intensity	high light intensity
0.00	0	0
0.02	20	33
0.04	29	53
0.06	35	68
0.08	39	79
0.10	42	86
0.12	45	89
0.14	46	90
0.16	46	90
0.18	46	90
0.20	46	90

   (a) Plot these results on a graph, drawing one line for low and one for high light intensity, both on the same pair of axes.
   (b) What is the CO_2 concentration of normal air?
   (c) What is the rate of photosynthesis at this CO_2 concentration in a high light intensity?
   (d) Market gardeners often add carbon dioxide to the air in greenhouses. What is the advantage of doing this?
   (e) Up to what values does CO_2 concentration act as a limiting factor at high light intensities?

6 Respiration

6.1 Respiration releases energy from food.

Every cell in every living organism needs energy. Cells get their energy from food. The energy is released from the food by a process called **respiration**.

Energy can be released from food by combining it with oxygen. This is called **oxidation**. You can try this yourself, in Practical 6.1.

6.2 Respiration oxidises sugars in stages.

Of course, when substances are oxidised in living cells, they do not burn with flames! If sugar was oxidised like this, the cells would get so hot that they would be killed. In a cell, during respiration, sugar is oxidised very gradually in a series of small, controlled reactions. The reactions are controlled by enzymes. The result, though, is similar. Sugar is combined with oxygen, releasing energy. Carbon dioxide and water are formed. This happens inside the mitochondria in a cell.

The process of respiration can be summarised like this:

sugar + oxygen ⟶ carbon dioxide + water + energy

The sugar which is normally used is glucose, $C_6H_{12}O_6$. The balanced equation for respiration is this:

$$C_6H_{12}O_6 + 6O_2 \longrightarrow 6CO_2 + 6H_2O + \text{energy}$$

The carbon dioxide and water are by-products. The reaction produces energy for the cell. Much of the energy that is released when the sugars are oxidised in respiration is used by the cell. It could be used for active transport, or for moving things around inside the cell, or for building up proteins. It could be used for cell division and growth. If the cell is a muscle cell, energy will be used for contraction. If it is a nerve cell, it will be used for transmitting nerve impulses. Some of the energy will be released in the form of heat. This is especially important in mammals and birds, which use the heat released in respiration to keep their body temperatures constant, even when the air or water around them is cold.

Questions

1. In respiration, sugar is oxidised. What does this mean?
2. What is special about the way that oxidation happens inside cells?

Practical 6.1 To show that peanuts release energy when they are oxidised

1. Set up your apparatus as shown in Fig 6.1. You will also need a thermometer.
2. Take the temperature of the water in the test tube, and record it.
3. Using the mounted needle, hold the peanut in the Bunsen flame. The heat from the flame will give the peanut enough energy for it to begin to combine with oxygen in the air, so it starts to burn. This reaction is called **oxidation**, or **combustion**.
4. Hold the burning peanut under the test tube of water until it stops burning.
5. Quickly take the temperature of the water again.

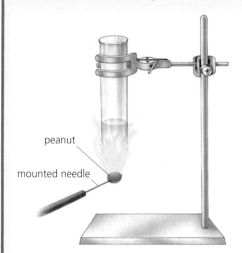

6.1 Burning a peanut.

Questions

1. Where did the energy come from (a) to start off the reaction of the peanut with oxygen in the air and (b) to raise the temperature of the water?
2. Why is it important to keep the Bunsen burner well away from the test tube?
3. Has *all* the energy in your peanut been released during this experiment? Give a reason for your answer.
4. Do you think all the energy that was released from the peanut went into the water? Explain your answer.

Practical 6.2 To show that heat is produced in respiration

1. Soak some peas in water for a day, so that they begin to germinate.
2. Boil a second set of peas, to kill them.
3. Wash both sets of peas in dilute disinfectant, so that any bacteria and fungi on them are killed.
4. Put each set of peas into a vacuum flask as shown in Fig 6.2. Do not fill the flasks completely.
5. Note the temperature of each flask.
6. Support each flask upside down, and leave them for a few days.
7. Note the temperature of each flask at the end of your experiment.

Questions

1. Which flask showed the higher temperature at the end of the experiment? Explain your answer.
2. Why is it important to kill any bacteria and fungi on the peas?
3. Why should the flasks not be completely filled with peas?
4. Carbon dioxide is a heavy gas. Why were the flasks left upside down, with porous cotton wool plugs in them?
5. Not all of the energy produced by the respiring peas will be given off as heat. What happens to the rest of it?

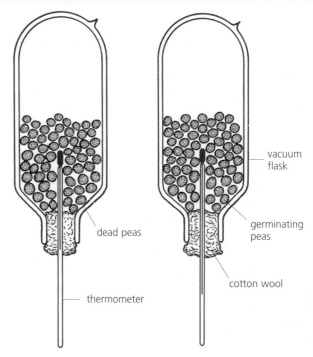

6.2 Apparatus to show that heat is produced during respiration.

6.3 Respiration sometimes occurs without oxygen.

The process described so far in this chapter releases energy from sugar by combining it with oxygen. It is called **aerobic respiration**, because it uses air (which contains oxygen).

It is possible, though, to release energy from sugar without using oxygen. It is not such an efficient process and not much energy is released, but the process is used by some organisms. It is called **anaerobic respiration** ('an' means without).

Yeast, a single-celled fungus, can respire anaerobically. It breaks down sugar to alcohol:

glucose ⟶ alcohol + carbon dioxide + energy

Some of the cells in your body, particularly muscle cells, can also respire anaerobically for a short time. They make lactic acid instead of alcohol:

glucose ⟶ lactic acid + energy

This is described in Section 6.12.

Table 6.1 A comparison of aerobic and anaerobic respiration.

Similarities
1. Energy released by breakdown of sugar

Differences

aerobic respiration	anaerobic respiration
1 Uses oxygen gas	Does not use oxygen gas
2 No alcohol or lactic acid made	Alcohol or lactic acid made
3 Large amount of energy released	Small amount of energy released

Questions

1. What is the purpose of respiration?
2. What is the energy released in respiration used for?
3. In which part of a cell does respiration take place?

Practical 6.3 To show that carbon dioxide is produced when yeast respires anaerobically

Lime water goes cloudy when carbon dioxide dissolves in it. Hydrogencarbonate indicator changes from red to yellow when carbon dioxide dissolves in it.

1. Boil some water, to drive off any dissolved air.
2. Dissolve a small amount of sugar in the boiled water, and allow it to cool.
3. When it is cool, add yeast and stir with a glass rod.
4. Set up the apparatus as in Fig 6.3. Add the liquid paraffin by trickling it gently down the side of the tube, using a pipette.
5. Set up an identical piece of apparatus, but use boiled yeast instead of living yeast.
6. Leave your apparatus in a warm place.
7. Observe what happens to the indicator solution after half an hour.

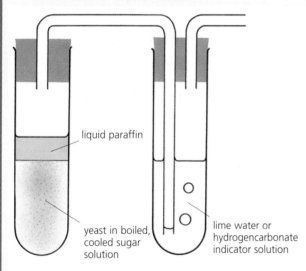

6.3 Apparatus to show that carbon dioxide is produced when yeast respires anaerobically.

Questions

1. Why is it important to boil the water?
2. Why must the sugar solution be cooled before adding the yeast?
3. What is the liquid paraffin for?
4. What happened to the lime water or hydrogencarbonate indicator solution in each of your pieces of apparatus? What does this show?
5. What new substance would you expect to find in the sugar solution containing living yeast at the end of the experiment?

Gaseous exchange

6.4 Gaseous exchange occurs at gas exchange surfaces.

If you look back at the respiration equation in Section 6.2, you will see that two substances are needed. They are glucose and oxygen. The way in which cells obtain glucose is described in Chapters 4 and 5. Animals get sugar from carbohydrates which they eat. Plants make theirs, by photosynthesis.

Oxygen is obtained in a different way. Animals and plants get their oxygen directly from their surroundings. The part of the organism through which oxygen enters the body is called the **gas exchange surface** (Table 6.2).

If you look again at the respiration equation you can see that carbon dioxide is made. This is a waste product and it must be removed from the organism. It leaves across the gas exchange surface (Table 6.3).

Table 6.2 Properties of gaseous exchange surfaces.

1. They should be **thin** to allow gases to diffuse across them quickly.
2. They should be close to an efficient **transport system** to take gases to and from the cells which need them.
3. They should be kept **moist**, to stop the cells on the surface from drying out and dying.
4. They should have a **large surface area** so that a lot of gas can diffuse across at the same time.
5. They should have a good **supply of oxygen**.

Gaseous exchange in humans

6.5 Gaseous exchange occurs in the respiratory system.

Fig 6.4 shows the structures which are involved in gaseous exchange in a human. The most important are the two lungs. Each lung is filled with many tiny air spaces called air sacs or **alveoli**. It is here that oxygen diffuses into the blood. Because they are so full of spaces, lungs feel very light and spongy to touch. The lungs are supplied with air through the windpipe or **trachea**.

6.6 Air is taken down into the lungs.

The nose and mouth

Air can enter the body through either the nose or mouth. The nose and mouth are separated by the **palate** (Fig 6.4), so you can breathe through your nose even when you are eating.

It is better to breathe through your nose, because the structure of the nose allows the air to become warm, moist and filtered before it gets to the lungs. Inside the nose are some thin bones called **turbinal bones** which are covered with a thin layer of cells. Some of these cells make a liquid containing water and mucus which evaporates into the air in the nose and moistens it (Fig 6.5).

Other cells have very tiny hair-like projections called **cilia**. The cilia are always moving and bacteria or particles of dust get trapped in them and in the mucus. Cilia are found all along the trachea and bronchi, too. They waft the mucus, containing bacteria and dust, up to the back of the throat, so that it does not block up the lungs.

The trachea

The air then passes into the windpipe or **trachea**. At the top of the trachea is a piece of cartilage called the **epiglottis**. This closes the trachea and stops food going down the trachea when you swallow. This is a reflex action, which happens automatically when a bolus of food touches the soft palate.

Just below the epiglottis is the voice box or **larynx**. This contains the vocal cords. The vocal cords can be tightened by muscles so that they make sounds when air passes over them. The trachea has rings of cartilage around it, which keep it open.

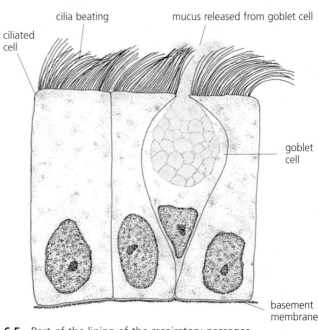

6.5 Part of the lining of the respiratory passages.

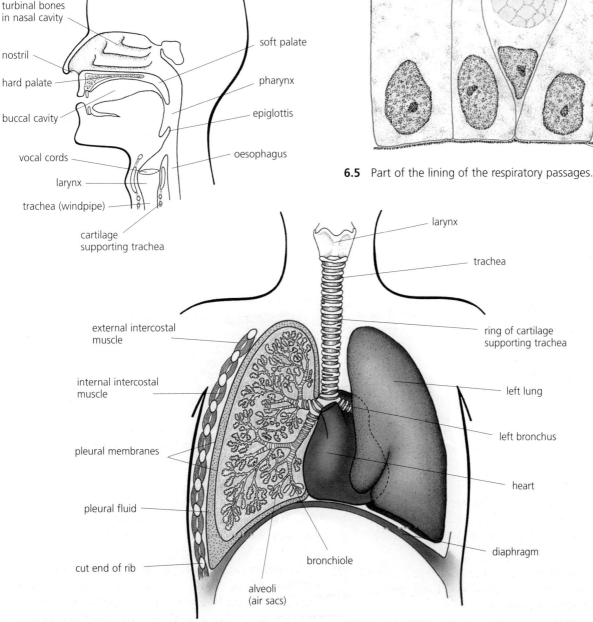

6.4 The human respiratory system.

The bronchi

The trachea goes down through the neck and into the **thorax**. The thorax is the upper part of your body from the neck down to the bottom of the ribs and diaphragm. In the thorax the trachea divides into two. The two branches are called the right and left **bronchi**. One bronchus goes to each lung and then branches out into many smaller tubes called **bronchioles**.

The alveoli

At the end of each bronchiole are tiny air sacs or **alveoli** (Fig 6.6). This is where gaseous exchange takes place.

6.7 Alveolar walls form the respiratory surface.

The walls of the alveoli are the respiratory surface. Tiny blood vessels, called **capillaries**, are closely wrapped around the outside of the alveoli (Fig 6.6). Oxygen diffuses across the walls of the alveoli into the blood (Fig 6.8). Carbon dioxide diffuses the other way.

The walls of the alveoli have several features which make them an efficient gaseous exchange surface (Table 6.2).

They are very thin.

Alveolar walls are only one cell thick. The capillary walls are also only one cell thick. An oxygen molecule only has to diffuse across this small thickness to get into the blood.

They have an excellent transport system.

Blood is constantly pumped to the lungs along the pulmonary artery. This branches into thousands of capillaries, which take blood to all parts of the lungs. Carbon dioxide in the blood can diffuse out into the air spaces in the alveoli and oxygen can diffuse into the blood. The blood is then taken back to the heart in the pulmonary vein, ready to be pumped to the rest of the body.

The way in which the blood carries oxygen and carbon dioxide is explained in Section 7.17.

They have a large surface area.

In fact, the surface area is enormous! The total surface area of all the alveoli in your lungs is over $100\,m^2$.

They have a good supply of oxygen.

Your breathing movements keep your lungs well supplied with oxygen.

Questions

1. Why is it better to breathe through your nose than through your mouth?
2. What is the function of the cilia in the respiratory passages?

Practical 6.4 Examining lungs

Examine some sheep lungs obtained from a butcher's shop or abattoir.

Questions

1. What colour are the lungs? Why are they this colour?
2. Push them gently with your finger. What do they feel like? Why do they feel like this?
3. What is covering the surface of the lungs? What is its name, and why is it there?
4. Find the two tubes leading down to the lungs. Which one is the oesophagus? Follow it along, and notice that it goes right past the lungs. Where is it going to?
5. The other tube is the trachea. What does it feel like? Why does it feel like this?
6. What is the name of the wide part at the top of the trachea? What is its function?
7. If the lungs have not been badly cut, take a long glass tube (such as a burette tube) and push it down through the trachea. Hold the trachea tightly against it, and blow down it. What happens?

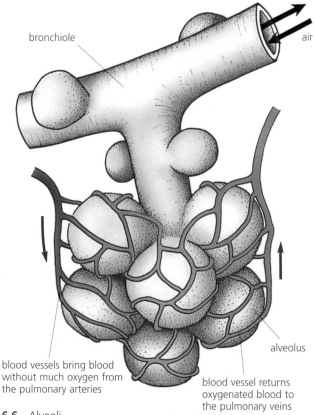

6.6 Alveoli.

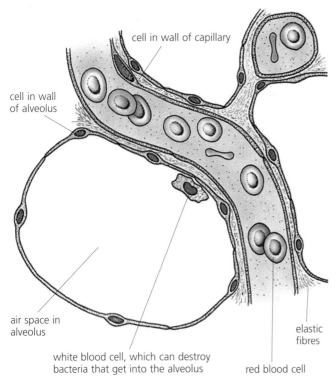

6.7 Section through part of a lung, magnified.

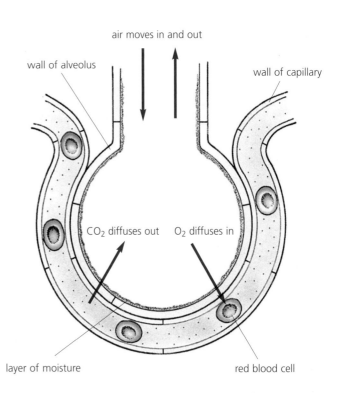

6.8 Gaseous exchange in an alveolus.

6.8 Ribs and diaphragm move during breathing.

To make air move in and out of the lungs, you must keep changing the volume of your thorax. First, you make it large so that air is sucked in. Then you make it smaller again so that air is squeezed out. This is called breathing or **ventilation**.

There are two sets of muscles which help you to breathe. One set is in between the ribs. This set is called the **intercostal muscles** made up of the **external** and **internal** intercostal muscles (Fig 6.9). The other set is in the **diaphragm**. The diaphragm is a large sheet of muscle and elastic tissue which stretches across your body, underneath the lungs and heart.

6.9 Breathing in is called inspiration.

When breathing in, the muscles of the diaphragm contract. This pulls the diaphragm downwards, which increases the volume in the thorax (Fig 6.10a, overleaf). At the same time, the external intercostal muscles contract. This pulls the rib cage upwards and outwards. Together, these movements increase the volume of the thorax.

As the volume of the thorax increases, the pressure inside it falls below atmospheric pressure. Extra space has been made and something must come in to fill it up. Air therefore rushes in along the trachea and bronchi into the lungs.

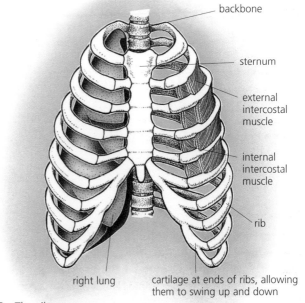

6.9 The rib cage.

Questions
1. What is the larynx?
2. What is the gaseous exchange surface of a human?

49

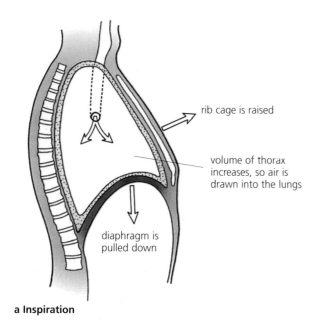

a Inspiration

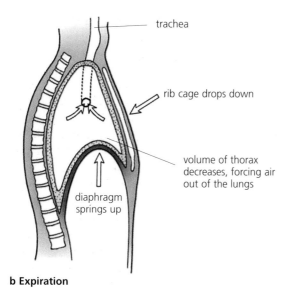

b Expiration

6.10 How the thorax changes shape during breathing.

6.10 Breathing out is called expiration.

When breathing out the muscles of the diaphragm relax. The diaphragm springs back up into its domed shape because it is made of elastic tissue. This decreases the volume in the thorax. The external intercostal muscles also relax. The rib cage drops down again into its normal position. This also decreases the volume of the thorax (Figs 6.10b and 6.11).

As the volume of the thorax decreases, the pressure inside it increases. Air is squeezed out through the trachea into the nose and mouth, and on out of the body.

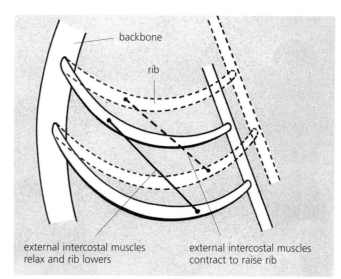

6.11 The external intercostal muscles raise and lower the ribs.

6.11 Internal intercostal muscles can force air out.

Usually, you breathe out by relaxing the external intercostal muscles and the muscles of the diaphragm, as explained in Section 6.10. Sometimes, though, you breathe out more forcefully – when coughing, for example. Then the internal intercostal muscles contract strongly, making the rib cage drop down even further. The muscles of the abdomen wall also contract, helping to squeeze extra air out of the thorax.

Table 6.3 Comparison of inspired and expired air.
Inspired air is the air you breathe in. Expired air is the air you breathe out.

	Inspired air	Expired air	Reason for differences
Oxygen	21%	16%	Oxygen is absorbed across gaseous exchange surface, then used by cells in respiration
Carbon dioxide	0.04%	4%	Carbon dioxide is made by cells as a waste product of respiration, and is released across the gaseous exchange surface
Argon and other inert gases	1%	1%	
Nitrogen	78%	78%	Nitrogen gas is not used by cells
Water content (humidity)	Variable	Always high	Gaseous exchange surfaces must be kept moist. Some of this moisture evaporates and is lost as air is breathed out
Temperature	Variable	Always high	Air is warmed as it passes through the respiratory passages

Practical 6.5 Comparing the carbon dioxide content of inspired and expired air

You can use either lime water or hydrogencarbonate indicator solution for this experiment. Lime water changes from clear to cloudy when carbon dioxide dissolves in it. Hydrogencarbonate indicator solution changes from red to yellow when carbon dioxide dissolves in it.

1. Set up the apparatus as in Fig 6.12.
2. Breathe in and out gently through the rubber tubing. Do not breathe too hard. Keep doing this until the liquid in one of the flasks changes colour.

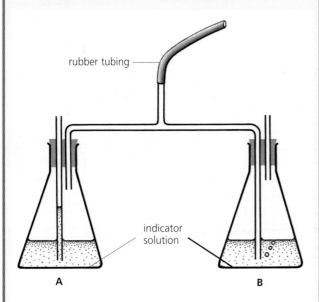

6.12 Comparing the carbon dioxide content of inspired and expired air.

Questions

1. In which flask did bubbles appear when you breathed out? Explain why.
2. In which flask did bubbles appear when you breathed in? Explain why.
3. What happened to the liquid in flask A?
4. What happened to the liquid in flask B?
5. What do your results tell you about the amount of carbon dioxide in inspired air and expired air?

6.12 Exercise can create an oxygen debt.

All the cells in your body need oxygen for respiration and all of this oxygen is supplied by the lungs. The oxygen is carried by the blood to every part of the body.

Sometimes, cells may need a lot of oxygen very quickly. Imagine you are running in a race. The muscles in your legs are using up a lot of energy. To produce this energy, the mitochondria in the muscles will be combining oxygen with glucose as fast as they can, to provide the energy for the muscles.

A lot of oxygen is needed to work as hard as this. You breathe deeper and faster to get more oxygen into your blood. Your heart beats faster to get the oxygen to the leg muscles as quickly as possible. Eventually a limit is reached. The heart and lungs cannot supply oxygen to the muscles any faster. But more energy is still needed for the race. How can that extra energy be found?

Extra energy can be produced by anaerobic respiration. Some glucose is broken down without combining it with oxygen:

$$\text{glucose} \longrightarrow \text{lactic acid} + \text{energy}$$

As explained in Section 6.3, this does not release very much energy, but a little extra might make all the difference.

When you stop running, you will have quite a lot of lactic acid in your muscles and your blood. This lactic acid must be broken down by combining it with oxygen. So, even though you do not need the energy any more, you go on breathing hard. You are taking in extra oxygen to break down the lactic acid.

While you were running, you built up an **oxygen debt**. You 'borrowed' some extra energy, without 'paying' for it with oxygen. Now, as the lactic acid is combined with oxygen, you are paying off the debt. Not until all the lactic acid has been used up, does your breathing rate and rate of heart beat return to normal.

Questions

1. Explain the difference between respiration and breathing.
2. Give two ways in which anaerobic respiration in yeast is similar to anaerobic respiration in humans.
3. Give one way in which anaerobic respiration in yeast differs from anaerobic respiration in humans.
4. Explain why, even when you have finished running a race, you still carry on breathing more quickly and deeply than usual for several minutes.

Practical 6.6 Investigating how breathing rate changes with exercise

1. Copy the table, ready to fill in your results.
2. Sit quietly for two minutes, to make sure you are completely relaxed.
3. Count how many breaths you take in one minute. Record it in your table.
4. Wait one minute, then count breaths again, and record.
5. Now do some vigorous exercise, such as stepping up and down onto a chair, for exactly two minutes. At the end of this time, sit down. Immediately count your breaths in the next minute, and record.
6. Continue to record your breaths per minute every other minute, until they have returned to near the level before you started to exercise.
7. Draw a graph of your results, putting time on the bottom (x) axis.

Results table

Time /minutes	Breathing rate /number of breaths per minute
1	
3	
6	
8	
10	

Questions

1. Why does your breathing rate rise so quickly during exercise?
2. Why did your breathing rate not go back to normal as soon as you finished exercising?
3. Work out how many minutes it took your breathing rate to return to normal after exercise. Collect everyone's results for this, and work out the class average.
   Now compare individual results with the average one. Do fit people who regularly play a lot of sport come above or below this average result? Try to explain this.
4. As well as **rate** of breathing, your **depth** of breathing will also have increased when you exercised. Suggest how you could measure this.

6.13 These athletes will have to pay back their oxygen debts at the end of the race.

Smoking

6.13 Cigarette smoking damages lungs and heart.

Cigarette smoke is very harmful to a person's lungs. Everyone who smokes does some damage to their lungs. Heavy smoking can also damage the heart and blood vessels.

Even people who do not smoke may be harmed by other people's cigarettes. If a baby's parents each smoke 20 cigarettes a day, the child will have indirectly smoked 80 cigarettes by the time it is one year old. Non-smokers are more likely to get coughs and bronchitis if they spend much time in a room with people who are smoking. In many places, cigarette smoking is no longer seen as socially acceptable.

Cigarette smoke contains three main ingredients. These are nicotine, tar and carbon monoxide (Fig 6.14). Each of these has its own effects on the body.

Table 6.4 The differences between respiration, gaseous exchange and breathing.

Respiration is a series of chemical reactions which happen in all living cells, in which food is broken down to release energy, usually by combining it with oxygen.

Gaseous exchange is the exchange of gases across a surface. For example, oxygen is taken into the body, and carbon dioxide is removed from it. Gaseous exchange also takes place during photosynthesis and respiration in plants.

Breathing (ventilation) is muscular movements which keep the gaseous exchange surface supplied with oxygen.

6.14 Nicotine is addictive.

Nicotine affects the brain. It is a stimulant, which means that it makes you feel more alert and active. It makes heart rate and blood pressure increase. People who smoke are more likely to suffer from heart disease than people who do not.

Nicotine is a very poisonous substance. It is used in some insecticides. A smoker takes in about 2 mg of nicotine for each cigarette that is smoked. The nicotine is absorbed into the blood. But fifteen minutes after smoking the cigarette, half of this nicotine has been broken down, so only 1 mg remains in the blood.

Nicotine is **addictive**. This means that once your body has got used to it, it is very hard to do without it. This is why many smokers find it difficult to give up smoking.

6.15 Tar increases the chances of getting lung cancer.

The tar in cigarette smoke is absorbed by some of the cells in the lungs, especially the ones lining the bronchi and bronchioles. Normally, these cells form a thin, protective layer (Fig 6.5). But the tar makes them divide, and build up into a thicker layer. Some of these groups of dividing cells may go on and on dividing, developing into **cancer**.

Non-smokers do occasionally get lung cancer, but this is very rare. Most deaths from lung cancer are of people who smoked and all forms of cancer are more common in smokers than non-smokers.

Tar is an irritant. It makes the linings of the respiratory passages inflamed, causing chronic bronchitis. It also damages the cilia lining these passages, and causes extra mucus to be made by the goblet cells. This mucus trickles down into the lungs – the cilia are no longer able to beat and sweep it upwards.

Bacteria breed in the mucus, causing infections. The person coughs, to try to move the mucus upwards. The constant coughing can damage the delicate alveoli in the lungs. This makes it much more difficult for the person to get enough oxygen into their blood. They have **emphysema**.

6.16 Carbon monoxide cuts down oxygen supply.

Carbon monoxide is absorbed into the blood. Here it combines rapidly with haemoglobin inside the red cells (Section 7.17). This means that there is less haemoglobin available to carry oxygen. In a heavy smoker who inhales, up to one fifth of his or her haemoglobin is combined with carbon monoxide instead of oxygen.

As well as cutting down the amount of oxygen *in* the blood, carbon monoxide also makes body cells less able to absorb oxygen *from* the blood. Brain cells are especially sensitive to this.

Babies born to mothers who smoke tend to be smaller than babies of non-smokers, and this is probably because of the effects of carbon monoxide on the baby in the womb.

6.17 Smoking will be the biggest killer.

If the World Health Organisation (WHO) forecasts are correct, smoking could become the world's biggest killer by the year 2020, causing more deaths than AIDS, tuberculosis, road accidents, murder and suicide put together.

Smoking-related diseases are killing 4 million people a year worldwide and that number will rise to 10 million a year in the next 25 years. Of these, 7 million deaths will occur in developing countries.

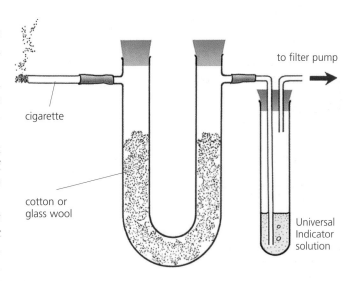

Smoke particles and tar are trapped in the cotton wool. The Universal Indicator solution changes from green (neutral) to yellow (acidic) because substances in the smoke form an acidic solution.

6.14 Apparatus to show what is contained in cigarette smoke.

Chapter revision questions

1. Describe an experiment you could do to see if germinating seeds give off carbon dioxide. Include a labelled diagram of your apparatus, and a control, and explain what you think your results would be.
2. Construct a table to compare the processes of respiration and photosynthesis in a green plant.

7 Transport

7.1 Large organisms need transport systems.

Large organisms need transport systems to supply all their cells with food, oxygen and other materials. This chapter describes the transport systems of mammals and flowering plants, and describes the parts which make up these systems.

Layout of a mammal transport system

7.2 Mammals have double circulatory systems.

The main transport system of a mammal is its blood system. This is sometimes called the vascular system. It is a network of tubes, called **blood vessels**. A pump, the **heart**, keeps blood flowing through the vessels.

Fig 7.1 illustrates the general layout of the blood system. The white arrows show the direction of blood flow. If you follow the arrows, beginning at the lungs, you can see that blood flows into the left-hand side of the heart, and then out to the rest of the body. It is brought back to the right-hand side of the heart, before going back to the lungs again.

This is called a **double circulatory system**, because the blood travels through the heart twice on one complete journey around the body.

7.3 Blood becomes oxygenated and deoxygenated.

The blood in the left-hand side of the heart has come from the lungs. It contains oxygen, which was picked up by the capillaries surrounding the alveoli. It is called **oxygenated blood**.

This oxygenated blood is then sent around the body. Some of the oxygen in it is taken up by the body cells, which need oxygen for respiration.

When this happens the blood becomes **deoxygenated**. The deoxygenated blood is brought back to the right-hand side of the heart. It then goes to the lungs, where it becomes oxygenated once more.

7.1 The general layout of the circulatory system of a human, as seen from the front.

> **Questions**
> 1 Why do large organisms need transport systems?
> 2 What is a double circulatory system?
> 3 What is oxygenated blood?
> 4 Where does blood become oxygenated?
> 5 Which side of the heart contains oxygenated blood?

The heart

7.4 Heart structure is related to function.

The function of the heart is to pump blood around the body. It is made of a special type of muscle called **cardiac muscle**. This muscle contracts and relaxes regularly, throughout life.

Fig 7.2 illustrates a section through a heart. It is divided into four chambers. The two upper chambers are called **atria**. The two lower chambers are ventricles. The chambers on the left-hand side are completely separated from the ones on the right-hand side by a **septum**.

If you look at Fig 7.1, you will see that blood flows into the heart at the top, into the atria. Both of the atria receive blood. The left atrium receives blood from the **pulmonary veins**, which come from the lungs. The right atrium receives blood from the rest of the body, arriving through the **venae cavae**.

From the atria, the blood flows into the ventricles. The ventricles then pump it out of the heart. The blood in the left ventricle is pumped into the **aorta**, which takes the blood around the body. The right ventricle pumps blood into the **pulmonary artery**, which takes it to the lungs.

The job of the ventricles is quite different from the job of the atria. The atria simply receive blood, either from the lungs or the body and supply it to the ventricles. The ventricles pump blood out of the heart and all round the body. To help them to do this, the ventricles have much thicker, more muscular walls than the atria.

There is also a difference in the thickness of the walls of the right and left ventricles. The right ventricle pumps blood to the lungs, which are very close to the heart. The left ventricle, however, pumps blood all around the body. The left ventricle has an especially thick wall of muscle to enable it to do this. The blood flowing to the lungs in the pulmonary artery is at a much lower pressure than the blood in the aorta.

7.5 Coronary arteries supply heart muscle.

Fig 7.3 shows that there are blood vessels on the outside of the heart. They are called the **coronary arteries**. These vessels supply blood to the heart muscles.

It may seem odd that this is necessary, when the heart is full of blood. However, the muscles of the heart are so thick that the food and oxygen in the blood inside the heart would not be able to diffuse to all the muscles quickly enough. The heart muscle needs a constant supply of food and oxygen, so that it can keep contracting and relaxing. The coronary artery supplies this.

If the coronary artery gets blocked, for example by a blood clot, the cardiac muscles run short of oxygen. They cannot contract, so the heart stops beating. This is called a heart attack or **cardiac arrest**.

> **Questions**
> 1. What kind of muscle is found in the heart?
> 2. Which parts of the heart receive blood from (a) the lungs, and (b) the body?
> 3. Which parts of the heart pump blood into (a) the pulmonary artery, and (b) the aorta?
> 4. Why do the ventricles have thicker walls than the atria?
> 5. Why does the left ventricle have a thicker wall than the right ventricle?
> 6. What is the function of the coronary artery?

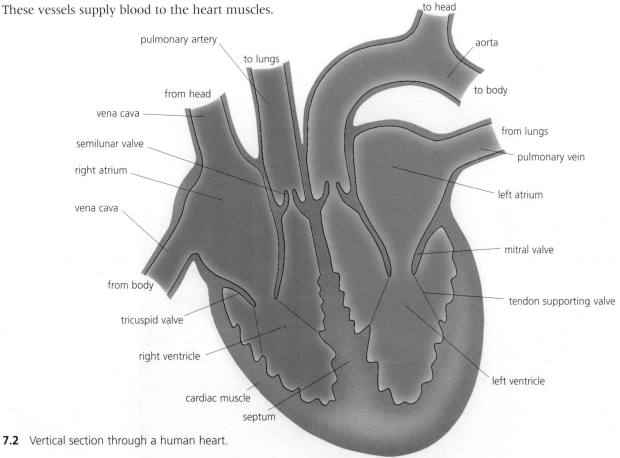

7.2 Vertical section through a human heart.

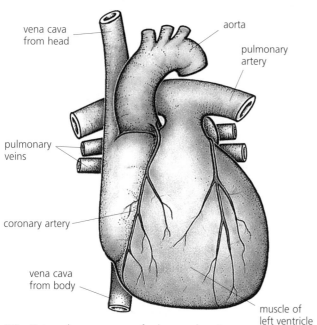

7.3 External appearance of a human heart.

7.6 The heart beats regularly.

The heart beats as the cardiac muscles in its walls contract and relax. When they contract, the heart becomes smaller, squeezing blood out. This is called **systole**. When they relax, the heart becomes larger, allowing blood to flow into the atria and ventricles. This is called **diastole**. Fig 7.4 illustrates this.

The rate at which the heart beats is controlled by a patch of muscle in the right atrium called the **pacemaker**. The pacemaker sends electrical signals through the walls of the heart at regular intervals, which make the muscle contract. The pacemaker's rate, and therefore the rate of heart beat, changes according to the needs of the body. For example, during exercise, when extra oxygen is needed by the muscles, the brain sends messages along nerves to the pacemaker, to make the heart beat faster.

Sometimes, the pacemaker stops working properly. An **artificial pacemaker** can then be placed in the person's heart. It produces an electrical impulse at a regular rate of about one impulse per second. Artificial pacemakers last for up to ten years before they have to be replaced.

7.7 Blood flows one way through heart valves.

There is a valve between the left atrium and the left ventricle, and another between the right atrium and ventricle. These are called **atrio-ventricular valves** (Fig 7.4).

Practical 7.1 To find the effect of exercise on the rate of heart beat

The best way to measure the rate of your heart beat is to take your pulse. Use the first two fingers of your right hand and lie them on the inside of your left wrist. Feel for the tendon near the outside of your wrist. If you rest your fingers lightly just over this tendon, you can feel the artery in your wrist pulsing as your heart pumps blood through it.

Perform the experiment as explained in Practical 6.6.

Questions

1. Why does your heart beat faster during exercise?
2. Why does the heart not return to its normal rate of beating as soon as you finish exercising?

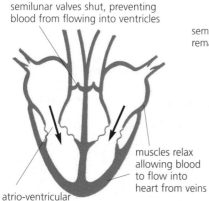

Diastole All muscles are relaxed. Blood flows into heart.

7.4 How the heart pumps blood.

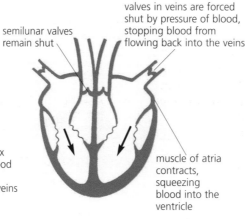

Atrial systole Muscles of atria contract. Muscles of ventricles remain relaxed. Blood forced from atria into ventricles.

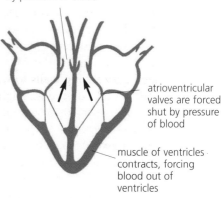

Ventricular systole Muscles of atria relax. Muscles of ventricles contract. Blood forced out of ventricles into arteries.

The valve on the left-hand side of the heart is made of two parts and is called the **bicuspid valve**, or the **mitral valve**. The valve on the right-hand side has three parts, and is called the **tricuspid valve**.

The function of these valves is to stop blood flowing from the ventricles back to the atria. This is important, so that when the ventricles contract, the blood is pushed up into the arteries, not back into the atria. As the ventricles contract, the pressure of the blood pushes the valves upwards. The tendons attached to them stop them from going up too far.

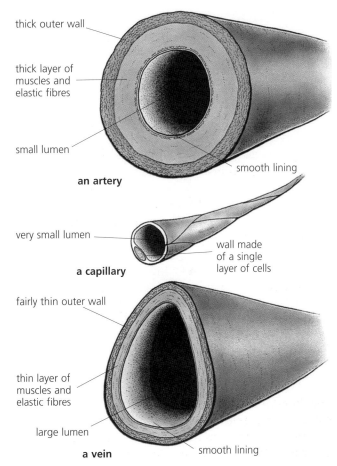

7.5 Sections through the three types of blood vessels.

Questions

1 What is (a) systole, and (b) diastole?
2 Where is the pacemaker, and what does it do?
3 Where are the atrio-ventricular valves?
4 What is their function?
5 Why are these valves supported by tendons?

Blood vessels

7.8 There are three kinds of blood vessels.

There are three main kinds of blood vessels: arteries, capillaries and veins. **Arteries** carry blood away from the heart. They divide again and again, and eventually form very tiny vessels called **capillaries**. The capillaries gradually join up with one another to form large vessels called **veins**. Veins carry blood towards the heart (Fig 7.5 and Table 7.1).

7.9 Arteries have thick elastic walls.

When blood flows out of the heart, it enters the arteries. The blood is then at very high pressure, because it has been forced out of the heart by the contraction of the muscular ventricles. Arteries therefore need very strong walls to withstand the high pressure of the blood flowing through them.

The blood does not flow smoothly through the arteries. It pulses through, as the ventricles contract and relax. The arteries have elastic tissue in their walls, which can stretch and recoil with the force of the blood. This helps to make the flow of blood smoother. You can feel your arteries stretch and recoil when you feel your pulse in your wrist.

The blood pressure in the arteries of your arm can be measured using a **sphygmomanometer**.

7.10 Capillaries are very narrow, with thin walls.

The arteries gradually divide to form smaller and smaller vessels (Figs 7.5 and 7.6). These are the capillaries. The capillaries are very small and penetrate to every part of the body. No cell is very far away from a capillary.

The function of the capillaries is to take food, oxygen and other materials to all the cells in the body, and to take away their waste materials. To do this, their walls must be very thin so that substances can get in and out of them easily. The walls of the smallest capillaries are only one cell thick (Fig 7.5).

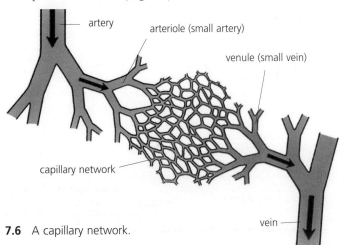

7.6 A capillary network.

7.11 Veins have one-way valves.

The capillaries gradually join up again to form veins. By the time the blood gets to the veins, it is at a much lower pressure than it was in the arteries. The blood flows more slowly and smoothly now. There is no need for veins to have such thick, strong, elastic walls.

If the veins were narrow, this would slow down the blood even more. To help to keep the blood moving easily through them, the space inside the veins, called the **lumen**, is much wider than the lumen of the arteries.

Veins have valves in them to stop the blood flowing backwards (Fig 7.7). Valves are not needed in the arteries, because the force of the heart beat keeps blood moving forwards through them.

Blood is also kept moving in the veins by the contraction of muscles around them. The large veins in your legs are squeezed by your leg muscles when you walk. This helps to push the blood back up to your heart. If a person is confined to bed for a long time, then there is a danger that the blood in these veins will not be kept moving. A clot may form in them, called a **thrombosis**. If the clot is carried to the lungs, it could get stuck in the arterioles of the lung. This is called a **pulmonary embolism**, and it may prevent the circulation reaching part of the lungs. In serious cases this can cause death.

7.12 Each organ has its own blood supply.

Figs 7.8 and 7.9 illustrate the positions of the main arteries and veins in the body.

Each organ of the body, except the lungs, is supplied with oxygenated blood from an artery. Deoxygenated blood is taken away by a vein. The artery and vein are named according to the organ they are connected with. For example, the blood vessels of the kidneys are the renal artery and vein. The liver has the hepatic artery and vein.

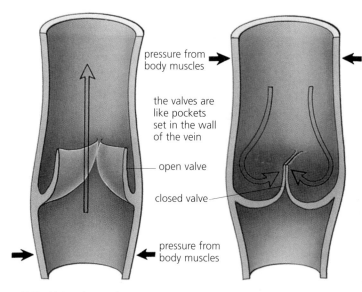

7.7 Valves in a vein.

All arteries, other than the pulmonary artery, branch from the aorta. All veins, except the pulmonary veins, join up to one of the two venae cavae.

The liver has two blood vessels supplying it with blood. The first is the **hepatic artery**, which supplies oxygen. The second is the **hepatic portal vein**. This vein brings blood from the digestive system (Fig 4.22), so that the liver can process the food which has been absorbed, before it travels to other parts of the body. All the blood leaves the liver in the **hepatic vein**.

> ### Questions
> 1 Which blood vessels carry blood (a) away from, and (b) towards the heart?
> 2 Why do arteries need strong walls?
> 3 Why do arteries have elastic walls?
> 4 What is the function of capillaries?
> 5 Why do veins have a large lumen?

Table 7.1 Arteries, veins and capillaries.

	Function	Structure of wall	Width of lumen	Reasons for structure
Arteries	Carry blood away from the heart	Thick, strong, containing muscles and elastic fibres	Varies, as elastic fibres stretch and recoil	Strength and elasticity needed to resist pulsing of blood as it is pumped by the heart
Capillaries	Supply all cells with their requirements, and take away waste products	Very thin, often only one cell thick	Very narrow, just wide enough for a red blood cell to squeeze through	No need for strong walls, as most of force of blood has been lost; thin walls and narrow lumen bring blood into close contact with tissues
Veins	Return blood to heart	Quite thin, containing far fewer muscles and elastic fibres than arteries	Wide, contain valves	No need for strong walls, as most of force of blood has been lost; wide lumen offers less resistance to blood flow; valves prevent backflow

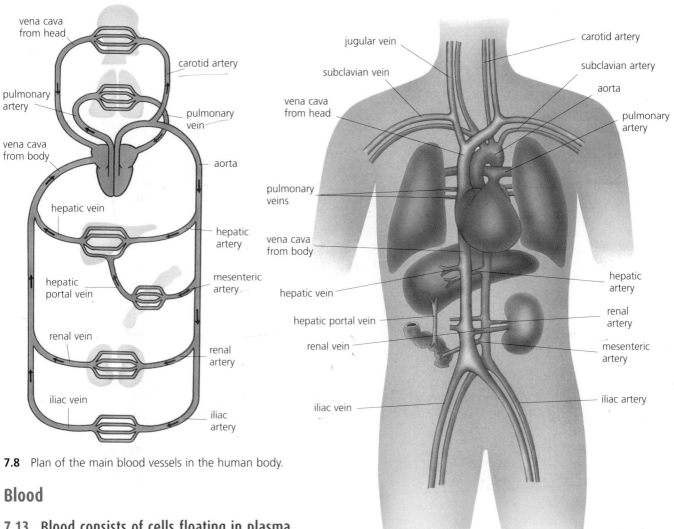

7.8 Plan of the main blood vessels in the human body.

7.9 The main arteries and veins in the human body.

Blood

7.13 Blood consists of cells floating in plasma.

The liquid part of blood is called **plasma**. Floating in the plasma are cells. Most of these are **red blood cells**. A much smaller number are **white blood cells**. There are also small fragments formed from special cells in the bone marrow, called **platelets** (Fig 7.10).

Plasma is mostly water. Many substances are dissolved in it. Glucose, amino acids, salts, hormones, blood proteins and antibodies are all dissolved in the plasma. These are described in Table 7.2.

7.14 Red blood cells carry oxygen.

Red blood cells are made in the bone marrow of some bones, including the ribs, vertebrae and some limb bones. They are produced at a very fast rate – about 9000 million per hour!

Red cells have to be made so quickly because they do not live for very long. Each red cell only lives for about four months. One reason for this is that they do not have a nucleus (Figs 7.10 and 7.11).

Red cells are red because they contain the pigment **haemoglobin**. This carries oxygen. Haemoglobin is a protein, and contains iron. The lack of a nucleus means that there is more space for haemoglobin.

Another unusual feature of red blood cells is their shape. They are biconcave discs – like a flat disc that has been pushed in on both sides. This, together with their very small size, gives them a relatively large surface area compared with their volume.

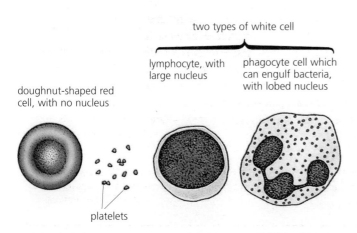

7.10 Blood cells.

This high surface area to volume ratio speeds up the rate at which oxygen can diffuse into and out of the cell. The small size of red blood cells is also useful in another way; it allows them to squeeze through the tiniest of blood capillaries, taking oxygen as close as possible to the cells that need it.

Old red blood cells are broken down in the liver, spleen and bone marrow. Some of the iron from the haemoglobin is stored, and used for making new haemoglobin. Some of it is turned into bile pigment and excreted.

7.15 White blood cells fight infection.

White cells are made in the bone marrow and in the **lymph nodes** (Section 7.21). White cells do have a nucleus, which is often quite large and lobed (Figs 7.10 and 7.11). They can move around, like *Amoeba*, and can squeeze out through the walls of blood capillaries into almost all parts of the body. Their function is to fight infection, and to clear up any dead body cells. This is described in Chapter 13.

7.16 Platelets help blood to clot.

Platelets are small fragments of cells, with no nucleus. They are made in the bone marrow, and they are involved in blood clotting.

Blood clotting stops pathogens entering through cuts. Normally, your skin provides a very effective

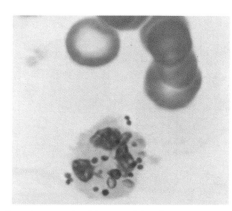

7.11 A light micrograph of human blood. You can see several red cells, and also a white cell, which is taking in and digesting bacteria.

barrier stopping harmful bacteria getting into your blood. However, if you cut yourself, there is a chance for them to get in through the cut. Blood clotting helps to stop this happening, as well as stopping you from losing too much blood (Fig 7.12).

Fig 7.13 summarises what happens when your blood clots. Platelets are very important in this process. Normally, blood vessel walls are very smooth. When a blood vessel is cut, the platelets bump into the rough edges of the cut, and react by releasing a chemical. The damaged tissues around the blood vessel also release chemicals.

Table 7.2 Some of the main components of blood plasma.

Substance	Source	Destination	Notes
Water	Absorbed in colon	All cells	Excess is removed by kidneys
Plasma proteins e.g. fibrinogen, antibodies	Fibrinogen made in the liver; antibodies made by lymphocytes	Remain in the blood	Fibrinogen helps in blood clotting; antibodies kill bacteria
Lipids including cholesterol and fatty acids	Absorbed in the ileum; also derived from fat reserves in the body	To the liver, for breakdown; to adipose tissue, for storage	Breakdown of fats yields energy; high cholesterol levels in the blood may increase the chances of heart disease
Carbohydrates e.g. glucose	Absorbed in the ileum; also produced by glycogen breakdown in the liver	To all cells, for energy release by respiration	Excess glucose is converted to glycogen and stored in liver and muscles
Excretory substances e.g. urea	Produced by amino acid deamination in the liver	To kidneys for excretion	
Mineral ions e.g. Na^+, Cl^-	Absorbed in the ileum and colon	To all cells.	Excess ions are excreted by the kidneys
Hormones	Secreted into the blood by endocrine glands	To all parts of the body.	Hormones only affect their own target organs; hormones are broken down in the liver, and they are also excreted by the kidneys
Dissolved gases e.g. carbon dioxide	Carbon dioxide is released from all cells as a waste product of respiration	To the lungs for excretion.	Most carbon dioxide is carried as hydrogencarbonate ions (HCO_3^-) in the plasma

In the blood plasma, there is a soluble protein called **fibrinogen**. The chemicals released by the platelets and the damaged tissues set off a chain of reactions, which cause the fibrinogen to change into **fibrin**.

Fibrin is insoluble. As its name suggests, it forms fibres. These fibres form a mesh across the wound. Red blood cells and platelets get trapped in the tangle of fibrin fibres, forming a **blood clot**.

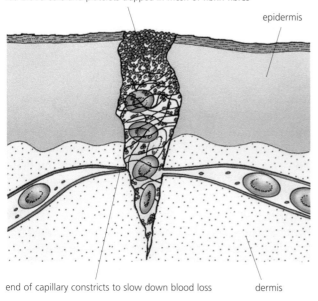

7.12 Vertical section through a blood clot.

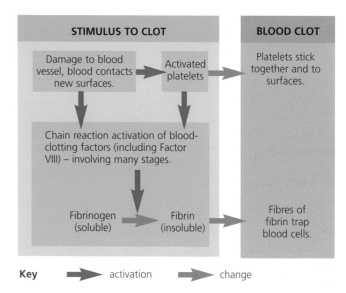

7.13 The sequence of events which occurs as blood clots.

> ### Questions
> 1 List five components of plasma.
> 2 Where are red blood cells made?
> 3 Explain how red blood cells are adapted for their function.
> 4 What is haemoglobin?
> 5 Where are white blood cells made?
> 6 What are platelets?

7.17 Many substances are transported by blood.

Transport of oxygen

In the lungs, oxygen diffuses from the alveoli into the blood. The oxygen diffuses into the red blood cells, where it combines with the haemoglobin (Hb) to form oxyhaemoglobin (oxyHb).

The blood is then taken to the heart in the pulmonary veins and pumped out of the heart in the aorta. Arteries branch from the aorta to supply all parts of the body with oxygenated blood. When it reaches a tissue which needs oxygen, the oxyHb gives up its oxygen, to become Hb again.

Because capillaries are so narrow, the oxyHb in the red blood cells is taken very close to the tissues which need the oxygen. The oxygen only has a very short distance to diffuse.

OxyHb is bright red, whereas Hb is purplish-red. The blood in arteries is therefore a brighter red colour than the blood in veins.

Transport of carbon dioxide

Carbon dioxide is made by all the cells in the body as they respire. The carbon dioxide diffuses through the walls of the capillaries, into the blood.

Most of the carbon dioxide is carried by the blood plasma in the form of hydrogencarbonate ions, HCO_3^-. A small amount is carried by Hb in the red cells.

Blood containing carbon dioxide is returned to the heart in the veins, and then to the lungs in the pulmonary arteries. The carbon dioxide diffuses out of the blood and is passed out of the body on expiration.

Transport of food materials

Digested food is absorbed in the ileum. It includes amino acids, fatty acids and glycerol, monosaccharides (such as glucose), water, vitamins and minerals. These all dissolve in the plasma in the blood capillaries in the villi.

These capillaries join up to form the hepatic portal vein. This takes the dissolved food to the liver. The liver processes the food and returns some of it to the blood. The food is then carried, dissolved in the blood plasma, to all parts of the body.

Transport of urea

Urea, a waste substance, is made in the liver. It dissolves in the blood plasma, and is carried to the kidneys. The kidneys excrete it in the urine.

Transport of hormones

Hormones are made in endocrine glands. The hormones dissolve in the blood plasma, and are transported all over the body.

Transport of heat

Some parts of the body, such as the liver, muscles and brown fat, make a lot of heat. The blood transports the heat to all parts of the body. This prevents the liver, muscles and brown fat becoming too hot, and helps to keep the rest of the body warm.

Lymph and tissue fluid

7.18 Tissue fluid is leaked plasma.

Capillaries leak! The cells in their walls do not fit together exactly, so there are small gaps between them. Plasma can therefore leak out from the blood.

White blood cells can also get through these gaps. They are able to move and can squeeze out of the capillaries. Red blood cells cannot get out as they cannot change their shape.

So plasma and white cells are continually leaking out of the blood capillaries. The fluid formed in this way is called **tissue fluid**. It surrounds all the cells in the body (Fig 7.14).

7.19 The functions of tissue fluid.

Tissue fluid is very important. It supplies cells with all their requirements. These requirements, such as oxygen and food materials, diffuse from the blood, through the tissue fluid, to the cells. Waste products, such as carbon dioxide, diffuse in the opposite direction.

The tissue fluid is the immediate environment of every cell in your body. It is easier for a cell to carry out its functions properly if its environment stays constant. For example this means it should stay at the same temperature, and at the same concentration.

Several organs in the body work to keep the composition and temperature of the blood constant, and therefore the tissue fluid as well. This process is called **homeostasis**, and is described in Chapter 10.

Questions

1. Why is blood in arteries a brighter red than the blood in veins?
2. Describe two ways in which carbon dioxide is carried in the blood.
3. Where does carbon dioxide
    (a) enter the blood
    (b) leave the blood?
4. Which vessel transports digested food to the liver?
5. How is urea transported?
6. Name two functions of blood other than transport.

Table 7.3 Components of blood.

Component	Structure	Functions
Plasma	Water, containing many substances in solution	• Liquid medium in which cells and platelets can float • Transports CO_2 in solution • Transports food materials in solution • Transports urea in solution • Transports hormones in solution • Transports heat • Transports substances needed for blood clotting • Transports antibodies
Red cells	Biconcave discs, with no nucleus, containing haemoglobin	• Transport oxygen • Transport small amount of CO_2
White cells	Variable shape, with nucleus	• Engulf and destroy bacteria (phagocytosis) • Make antibodies
Platelets	Small particles, with no nucleus	• Help in blood clotting

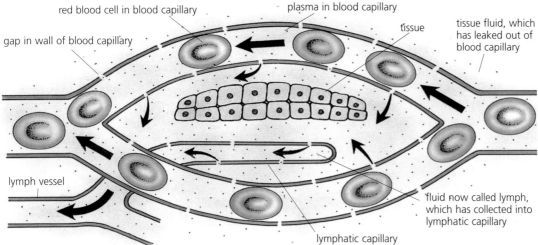

7.14 Part of a capillary network, to show how tissue fluid and lymph are formed.

7.20 Lymph is drained tissue fluid.

The plasma and white cells which leak out of the blood capillaries must eventually be returned to the blood. In the tissues, as well as blood capillaries, are other small vessels. They are **lymphatic capillaries** (Figs 7.14 and 7.15). The tissue fluid slowly drains into them. The fluid is now called **lymph**.

The lymphatic capillaries gradually join up to form larger **lymphatic vessels**. These carry the lymph to the subclavian veins, which bring blood back from the arms (Fig 7.16). Here the lymph enters the blood again.

The lymphatic system has no pump to make the lymph flow. Lymph vessels do have valves in them, however, to make sure that movement is only in one direction. Lymph flows much more slowly than blood.

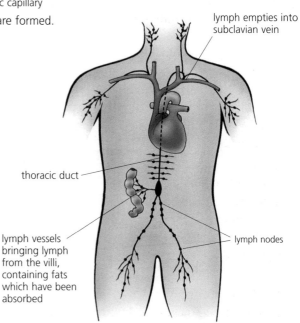

7.16 The main lymph vessels and lymph nodes.

7.21 Lymph nodes contain white blood cells.

On its way from the tissues to the subclavian vein, lymph flows through several **lymph nodes**. Some of these are shown in Fig 7.16.

Lymph nodes contain large numbers of white cells. Most bacteria or toxins (Section 13.3) in the lymph can be destroyed by these cells.

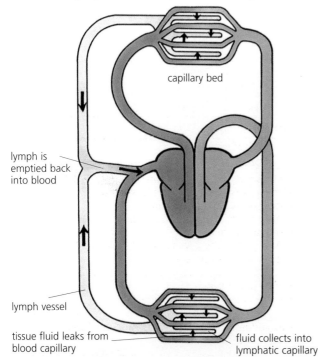

7.15 Relationship between blood circulation and lymphatic circulation.

> ### Questions
> 1 What is tissue fluid?
> 2 Give two functions of tissue fluid.
> 3 What is lymph?
> 4 Why do lymphatic capillaries have valves in them?
> 5 Name two places where lymph nodes are found.

Transport in a flowering plant

7.22 Plants have two transport systems – phloem and xylem.

Transport systems in plants are less elaborate than in mammals. Plants are less active than mammals, and so their cells do not need to be supplied with materials so quickly. Also, the branching shape of a plant means that all the cells can get their oxygen for respiration, and carbon dioxide for photosynthesis, directly from the air, by diffusion.

Plants have two transport systems. The **xylem vessels** carry water and minerals, while the **phloem tubes** carry food materials which the plant has made.

7.23 Xylem helps to support plants.

A xylem vessel is like a long drainpipe (Fig 7.17). It is made of many hollow, dead cells, joined end to end. The end walls of the cells have disappeared, so a long, open tube is formed. Xylem vessels run from the roots of the plant, right up through the stem. They branch out into every leaf.

Xylem vessels contain no cytoplasm or nuclei. Their walls are made of cellulose and **lignin**. Lignin is very strong, so xylem vessels help to keep plants upright. Wood is made almost entirely of lignified xylem vessels.

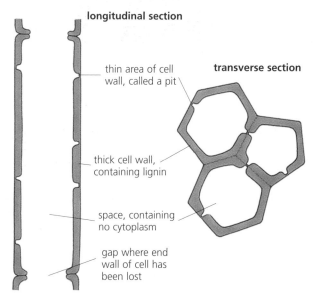

7.17 Xylem vessels.

7.24 Phloem contains sieve tube elements.

Like xylem vessels, phloem tubes are made of many cells joined end to end. However, their end walls have not completely broken down. Instead, they form **sieve plates** (Fig 7.18), which have small holes in them. The cells are called **sieve tube elements**. Sieve tube elements contain cytoplasm, but no nucleus. They do not have lignin in their cell walls.

Each sieve tube element has a **companion cell** next to it. The companion cell does have a nucleus, and also contains many other organelles. Companion cells probably supply sieve tube elements with some of their requirements.

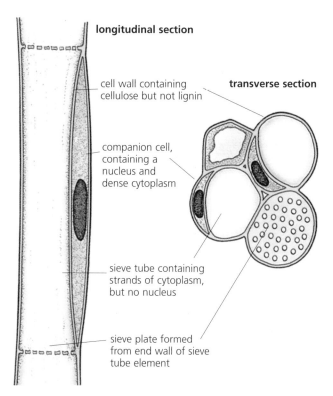

7.18 Phloem tubes.

7.25 Vascular bundles contain xylem and phloem.

Xylem vessels and phloem tubes are usually found close together. A group of xylem vessels and phloem tubes is called a **vascular bundle**.

The positions of vascular bundles in roots and shoots are shown in Figs 7.19 and 7.20. In a root, vascular tissue is found at the centre, whereas in a shoot vascular bundles are arranged in a ring near the outside edge. They help to support the plant.

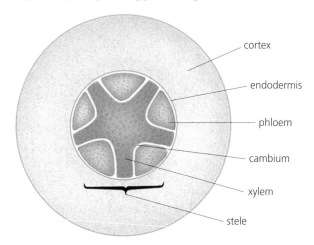

7.19 Transverse section of a root.

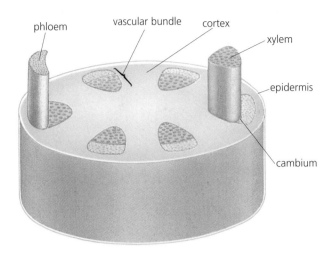

7.20 Transverse section of a stem.

Questions

1. Why do plants not need such elaborate transport systems as mammals?
2. What do xylem vessels carry?
3. What do phloem tubes carry?
4. What substance makes up the cell walls of xylem vessels?
5. Give three ways in which phloem tubes differ from xylem vessels.
6. What is a vascular bundle?
7. Think back to your work on leaf structure. Where in a leaf do you find xylem and phloem tissues?
8. Describe one function of xylem, other than transport of water and minerals.

Transport of water

7.26 Root structure is related to function.

Plants take in water from the soil, through their **root hairs**. The water is carried in the xylem vessels to all parts of the plant.

Fig 7.21 shows the structure of a root. At the very tip is the **root cap**. This is a layer of cells which protects the root as it grows through the soil. The rest of the root is covered by a layer of cells called the epidermis.

The **root hairs** are a little way up from the root tip. Each root hair is a long epidermal cell (Figs 7.21 and 7.22). Root hairs do not live for very long. As the root grows, they are replaced by new ones.

7.27 Root hairs absorb water by osmosis.

The function of a root hair is to absorb water and minerals from the soil. Water gets into a root hair by osmosis. The cytoplasm and cell sap inside it are quite concentrated solutions. The water in the soil is normally a dilute solution. Water therefore diffuses into the root hair, down its concentration gradient, through the partially permeable cell membrane. The large surface area of each root hair cell helps a lot of water to be absorbed very quickly.

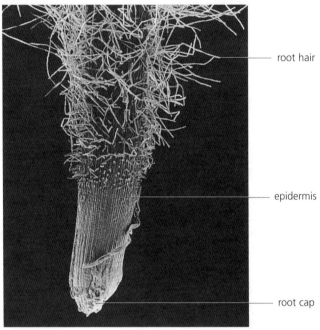

7.21 A root tip, showing the root cap and root hairs.

7.28 Absorbed water enters the xylem.

The root hairs are on the edge of the root. The xylem vessels are in the centre. Before the water can be taken to the rest of the plant, it must travel to these xylem vessels.

The path it takes is shown in Fig 7.22, overleaf. It travels by osmosis through the cortex, from cell to cell. Some of it also travels through the spaces between the cells.

7.29 Water is sucked up the xylem.

Water moves up xylem vessels in the same way that a drink moves up a straw when you suck it. When you suck a straw, you are reducing the pressure at the top of the straw. The liquid at the bottom of the straw is at a higher pressure, so it flows up the straw into your mouth.

The same thing happens with the water in xylem vessels. The pressure at the top of the vessels is lowered, while the pressure at the bottom stays high. Water therefore flows up the xylem vessels.

How is the pressure at the top of the xylem vessels reduced? It happens because of transpiration.

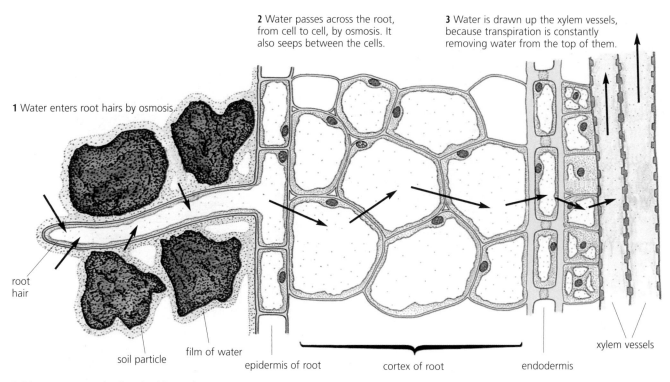

7.22 How water is absorbed by a plant.

Practical 7.2 To see which part of a stem transports water and solutes

1. Take a small plant with a root system intact. Wash the roots thoroughly.
2. Put the roots of the plant into eosin solution. Leave overnight.
3. Set up a microscope.
4. Remove the plant from the eosin solution, and wash the roots thoroughly.
5. Use a razor blade to cut across the stem of the plant about half way up. Take great care when using a razor blade and do not touch its edges.
6. Now cut very thin sections across the stem. Try to get them so thin that you can see through them. It does not matter if your section is not a complete circle.
7. Choose your thinnest section, and mount it in a drop of water on a microscope slide. Cover with a coverslip.
8. Observe the section under a microscope. Compare what you can see with Fig 7.20. Make a labelled drawing of your section.

Questions
1. Which part of the stem contained the dye? What does this tell you about the transport of water and solutes (substances dissolved in water) up a stem?
2. Why was it important to wash the roots of the plant:
   (a) before putting it into the eosin solution
   (b) before cutting sections?
3. Design an experiment to find out how quickly the dye is transported up the stem.

Questions

1. In which part of a root are root hairs found?
2. Suggest why root hairs are not found at the very tip of a root.
3. Explain why water moves into root hair cells from the soil.
4. How does water travel from a root hair cell to a xylem vessel?
5. Why does water move up xylem vessels?

7.30 Transpiration is evaporation from leaves.

Transpiration is the loss of water vapour from a plant. Most of this takes place from the leaves.

If you look back at Fig 5.4, you will see that there are openings on the underside of the leaf called **stomata**. The cells inside the leaf are each covered with a thin film of moisture.

Some of this film of moisture evaporates from the cells, and the water vapour diffuses out of the leaf through the stomata. Water from the xylem vessels in the leaf will travel to the cells by osmosis to replace it.

Water is constantly being taken from the top of the xylem vessels, to supply the cells in the leaves. This reduces the effective pressure at the top of the xylem vessels, so that water flows up them. This process is known as the **transpiration stream** (Fig 7.23).

7.31 Water moves down a water potential gradient.

You can think of the way that water moves into a root hair, across to the xylem vessels, up to the leaves and then out into the air in terms of water potential.

You will remember that water moves down a water potential gradient, from a high water potential to a low water potential (Section 2.3). All along this pathway, the water is moving down a water potential gradient from one place to another. The highest water potential is in the solution in the soil, and the lowest water potential is in the air.

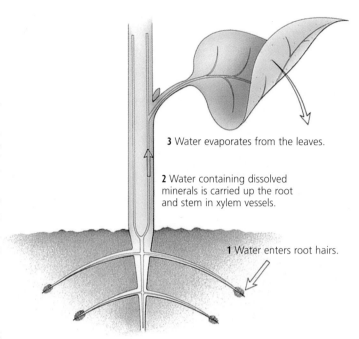

7.23 The transpiration stream.

7.32 A potometer compares transpiration rates.

It is not easy to measure how much water is lost from the leaves of a plant. It is much easier to measure how fast the plant takes up water. The rate at which a plant takes up water depends on the rate of transpiration – the faster a plant transpires, the faster it takes up water.

Fig 7.24 illustrates apparatus which can be used to compare the rates of transpiration in different conditions. It is called a **potometer**.

By recording how fast the air/water meniscus moves along the capillary tube, you can compare how fast the plant takes up water in different conditions.

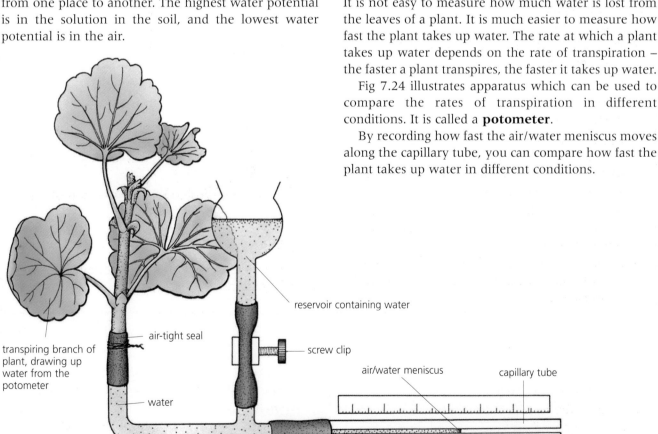

7.24 A potometer.

Practical 7.3 To see which surface of a leaf loses most water.

Cobalt chloride paper is blue when dry and pink when wet. Use forceps to handle it.

1 Use a healthy, well-watered potted plant, with leaves which are not too hairy. Fix a small square of blue cobalt chloride paper onto each surface of one leaf, using clear sticky tape. Make sure there are no air spaces around the paper.
2 Leave the paper on the leaf for a few minutes.

Questions
1 Which piece of cobalt chloride paper turned pink first?
  What does this tell you about the loss of water from a leaf?
2 Why does this surface lose water faster than the other?
3 Why is it important to use forceps, not fingers, for handling cobalt chloride paper?

Practical 7.4 To measure the rate of transpiration of a potted plant

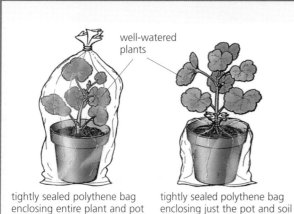

tightly sealed polythene bag enclosing entire plant and pot

tightly sealed polythene bag enclosing just the pot and soil

7.25 Measuring the rate of transpiration of a potted plant.

1 Use two similar well-watered potted plants. Enclose one plant entirely in a polythene bag, including its pot (Fig 7.25). This is the control.
2 Enclose only the pot of the second plant in a polythene bag. Fix the bag firmly around the stem of the plant, and seal with petroleum jelly.
3 Place both plants on balances, and record their masses.
4 Record the mass of each plant every day, at the same time, for at least a week.
5 Draw a graph of your results.

Questions
1 Which plant lost mass? Why?
2 Do you think this is an accurate method of measuring transpiration rate? How could it be improved?

Practical 7.5 Using a potometer to compare rates of transpiration under different conditions

1 Set up the potometer as in Fig 7.24. The stem of the plant must fit exactly into the rubber tubing, with no air gaps. Petroleum jelly will help to make an air-tight seal.
2 Fill the apparatus with water, by opening the screw clip.
3 Close the clip again, and leave the apparatus in a light, airy place. As the plant transpires, the water it loses is replaced by water taken up the stem. Air will be drawn in at the end of the capillary tube.
4 When the air/water meniscus reaches the scale, begin to record the position of the meniscus every two minutes.
5 When the meniscus reaches the end of the scale, refill the apparatus with water from the reservoir as before.
6 Now repeat the experiment, but with the apparatus in a different situation. You could try each of these: (a) blowing it with a fan, (b) putting it in a cupboard, (c) putting it in a refrigerator.
7 Draw a graph of your results.

Questions
1 Under which conditions did the plant transpire (a) most quickly, and (b) most slowly?
2 You have been using the potometer to compare the rate of uptake of water under different conditions. Does this really give you a good measurement of the rate of transpiration? Explain your answer.

7.33 Conditions affect transpiration rate.

Temperature
On a hot day, water evaporates quickly from the leaves of a plant. Transpiration increases as temperature increases.

Humidity
Humidity is the moisture content of the air. The higher the humidity, the less water will evaporate from the leaves. Transpiration decreases as humidity increases, because the diffusion gradient from the leaf to the air is less.

Wind speed
On a windy day, water evaporates more quickly than on a still day. Transpiration increases as wind speed increases.

Light intensity
In bright sunlight, a plant may open its stomata to supply plenty of carbon dioxide for photosynthesis. More water can therefore evaporate from the leaves.

Water supply
If water is in short supply, then the plant will close its stomata. This will cut down the rate of transpiration. Transpiration decreases when water supply decreases below a certain level.

7.34 Many plants can cut down water loss.

Transpiration is useful to plants, because it keeps water moving up the xylem vessels. But if the leaves lose too much water, the roots may not be able to take up enough to replace it. If this happens, the plant **wilts**. To stop this happening, many plants have ways of cutting down the rate of transpiration.

Closing stomata
Plants lose most water through their stomata. If they close their stomata, then transpiration will slow right down. Fig 7.27 shows how they do this.

However, if its stomata are closed, then the plant cannot photosynthesise, because carbon dioxide cannot diffuse into the leaf. Plants only close their stomata when they really need to, such as when it is very hot and dry, or when they could not photosynthesise anyway, such as at night.

Waxy cuticle
Many leaves, such as holly leaves, are covered with a waxy cuticle, made by the cells in the epidermis. The wax waterproofs the leaf.

Hairy leaves
Some plants have hairs on their leaves. These hairs trap a layer of moist air next to the leaf.

Stomata on underside of leaves
In most leaves, there are more stomata on the lower surface than on the upper surface. The lower surface is usually cooler than the upper one, so less water will evaporate.

Cutting down on the surface area
The smaller the surface area of the leaf, the less water will evaporate from it. Plants like cacti (Fig 7.26) have leaves (the spines) with a small surface area, to help them to conserve water. However, this slows down photosynthesis, because it means less light and carbon dioxide can be absorbed.

7.26 Ferocactus – a plant adapted to live in deserts.

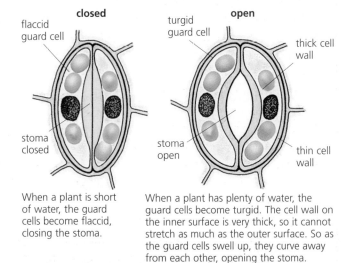

When a plant is short of water, the guard cells become flaccid, closing the stoma.

When a plant has plenty of water, the guard cells become turgid. The cell wall on the inner surface is very thick, so it cannot stretch as much as the outer surface. So as the guard cells swell up, they curve away from each other, opening the stoma.

7.27 How stomata open and close.

Uptake of minerals

7.35 Root hairs absorb minerals by active transport.

As well as absorbing water by osmosis, root hairs absorb mineral salts. These are in the form of ions dissolved in the water in the soil. They travel to the xylem vessels along with the water which is absorbed, and are transported to all parts of the plant.

These minerals are usually present in the soil in quite low concentrations. The concentration inside the root hairs is higher. In this situation, the mineral ions would normally diffuse out of the root hair into the soil. Root hairs can, however, take up mineral salts against their concentration gradient. It is the cell membrane which does this. Special carrier molecules in the cell membrane of the root hair carry the mineral ions across the cell membrane into the cell, against their concentration gradient.

This is called **active transport** (Section 2.9). It uses a lot of energy. The energy is supplied by respiration in the root hair cells.

Transport of manufactured food

7.36 Phloem translocates organic foods.

Leaves make carbohydrates by photosynthesis. They also use some of these carbohydrates to make amino acids, proteins, oils and other organic substances.

Some of the organic food material, especially sugar, that the plant makes is transported in the phloem tubes. It is carried from the leaves to whichever part of the plant needs it. This is called **translocation**. The sap inside phloem tubes therefore contains a lot of sugar, particularly sucrose. Energy is needed to make sap move through phloem tubes. It is an active process.

7.37 Systemic pesticides are translocated in phloem.

People who grow crops for food sometimes need to use chemicals called pesticides. Pests, such as insects that eat the crop plants, or fungi that grow on them, can greatly reduce the yield of the crop. Pesticides are used to kill the insects or fungi.

Some pesticides kill only the insects or fungus that the spray touches. They are called **contact pesticides**. They can be very effective if they are applied properly, but sometimes is is better to use a **systemic pesticide**. A systemic pesticide is absorbed into the plant and carried all the way through it in its phloem tissue. So any insect feeding on the plant, even if it was hidden under the leaf where the spray could not reach it, will eventually end up feeding on pesticide. The same is true for fungi; no matter where they are growing on the plant, the pesticide will eventually reach them. You can read more about this in Sections 16.17 and 16.18.

Chapter revision questions

1. Using Fig 7.8 to help you, list in order the blood vessels and parts of the heart which:

   (a) a glucose molecule would travel through on its way from your digestive system to a muscle in your leg
   (b) a carbon dioxide molecule would travel through on its way from the leg muscle to your lungs.

2. Explain the difference between each of the following pairs.

   (a) blood, lymph
   (b) diastole, systole
   (c) artery, vein
   (d) deoxygenated blood, oxygenated blood
   (e) atrium, ventricle
   (f) hepatic vein, hepatic portal vein
   (g) red blood cell, white blood cell
   (h) xylem, phloem
   (i) diffusion, active transport

3. Arteries, veins, capillaries, xylem vessels and phloem tubes are all tubes used for transporting substances in mammals and flowering plants. Describe how each of these tubes is adapted for its particular function.

4. (a) What is meant by a double circulatory system?
   (b) In an unborn child, the lungs do not work. The baby gets its oxygen from the mother, to which it is connected by the umbilical cord. This cord contains a vein, which carries the oxygenated blood to the baby's vena cava.

   (i) Which chamber of the heart does oxygenated blood enter in an adult person?

   (ii) Which chamber of the heart does oxygenated blood enter in an unborn baby?

   (iii) In an unborn child, there is a hole in the septum between the left and right atria. What purpose do you think this has?

   (iv) As soon as the baby takes its first breath, this hole closes up. Why is this important?

8 Reproduction

8.1 Reproduction may be sexual or asexual.

Living organisms may be killed by other organisms, or die of old age. New organisms have to be produced to replace those that die. This is reproduction.

Each species reproduces in a different way. However, there are only two basic types – **asexual reproduction**, and **sexual reproduction**.

Asexual reproduction

8.2 Growth and asexual reproduction involve mitosis.

A living organism such as a person is made of many cells. You began life as a single cell. This cell grew, and divided to form two cells. These cells grew, and divided again to form four cells. This process went on and on until all the cells of which you are now made had been produced.

The way in which cells divide during growth is shown in Fig 8.1. It is called **mitosis**. Many living organisms also use mitosis to make complete new organisms – they just grow a new one out of themselves. This is called **asexual reproduction.**

8.3 Mitosis makes new cells with the same genes as the old one.

The nucleus of a cell contains threads of **DNA**, called **chromosomes** (Fig 8.2). Different species of organisms have different numbers of chromosomes. Human cells, for example, have 46 chromosomes. The DNA is a coded recipe for making proteins. Each chromosome contains many recipes, or **genes**. You can find out more about DNA, genes and chromosomes in Chapter 12.

Mitosis is a method of cell division which makes two new cells with exactly the same number of chromosomes, carrying exactly the same genes, as the old cell.

8.4 Asexual reproduction produces genetically identical offspring.

As asexual reproduction takes place by means of mitosis, the new organisms produced are genetically identical to their parent, and to each other. A group of genetically identical organisms is called a **clone**.

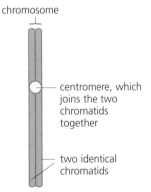

8.2 A chromosome just before division.

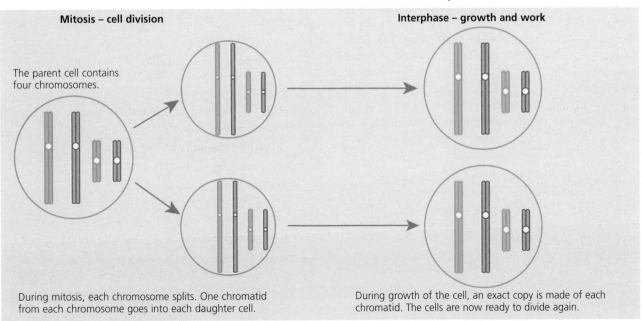

8.1 Chromosomes during the life of a cell dividing by mitosis.

8.5 Bacteria reproduce by binary fission.

Bacteria are tiny organisms made of a single cells. Unlike the cells of animals and plants, they do not have a nucleus containing chromosomes. They therefore cannot divide by mitosis.

The DNA of a bacterium is just a single, circular molecule. When the bacterium is about to divide, an exact copy is made of this DNA. The cell then divides into two, with one DNA molecule in each half. Two new cells have therefore been made, each genetically identical to the original cell. This method of asexual reproduction is called **binary fission**. It is shown in Fig 8.3.

When they are in suitable conditions, at the right temperature and with plenty of nutrients and water, bacteria can reproduce very quickly. One cell may be able to divide after only 20 minutes. So, if you start with one cell, there will be two after 20 minutes, four after 40 minutes, eight after one hour and so on. After five hours, 32 768 genetically identical bacteria will have been formed from just one original parent cell. You might like to try to calculate how many there will be after 24 hours!

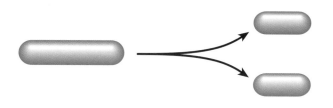

8.3 Binary fission in bacteria.

8.6 Fungi produce spores asexually.

Fungi are living organisms that are neither animals, plants or bacteria. Like plants, their cells have cell walls, but these walls often contain a substance called **chitin**, not cellulose. Like animals, they cannot photosynthesise, and they feed on organic food materials. Fig 8.4 shows a fungus called *Mucor*, or bread mould, which often grows and feeds on bread.

You may be able to grow some *Mucor* if you leave a piece of damp bread lying around for a few days; put a loose cover over it to stop it drying out. The fungus looks like a furry growth on the bread. This 'fur' is called a **mycelium**, and it is made up of many threads called **hyphae** (singular: **hypha**). Each hypha is just one cell thick. Some of the hyphae grow through the bread, where they secrete enzymes which digest the starch, protein and fat in the bread to produce glucose, amino acids, fatty acids and glycerol. These substances then diffuse into the hypha.

Other hyphae grow upwards, to form **aerial hyphae**. A swelling called a **sporangium** forms at the tip of each aerial hypha, and inside it cells divide asexually to form many genetically identical **spores**. Each spore is surrounded by a hard, protective coat that stops it from drying out. The spores are very small and light, and can be blown away on air currents, or carried away stuck to the feet of a housefly. Some of them may land on another suitable piece of food, where they will germinate and grow into a new mycelium.

8.7 Plants can be propagated asexually.

Many plants are able to reproduce asexually, and gardeners and farmers make use of this. Asexual reproduction can quickly and efficiently produce many new plants, all genetically identical to one another. This is advantageous if the original plant had exactly the characteristics that are wanted, such as large and attractive flowers, or good flavour, or high yield.

Tuber formation in potatoes

Potatoes, for example, reproduce using **stem tubers** (Fig 8.5). Some of the plant's stems grow normally, above ground, producing leaves, which photosynthesise. Other stems grow under the soil. Swellings called tubers form on them. Sucrose is transported from the leaves into these underground stem tubers, where it is converted into starch and stored. The tubers grow larger and larger. Each plant can produce many stem tubers.

The tubers are harvested, to be used as food. Some of them, however, are saved to produce next year's crop. These tubers are planted underground, where they grow shoots and roots to form a new plant. Because each potato plant produces many tubers, one plant can give rise to many new ones. To get more plants, tubers can be cut into several pieces. As long each piece has a bud on it, it can grow into a complete new plant.

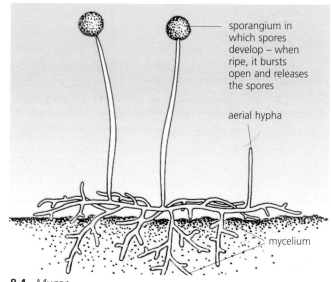

8.4 *Mucor.*

8.5 Tuber formation in potatoes.

Taking cuttings

Another commercially important way of producing many genetically identical plants is by taking **cuttings** (Fig 8.6). Many plants that people like to grow in their houses, for decoration, are propagated in this way. So are some important crops such as bananas.

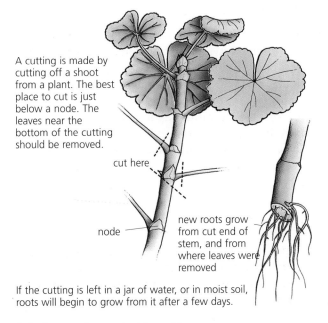

8.6 Propagating a plant by cuttings.

Tissue culture

Yet another method of producing large numbers of genetically identical plants uses a technique called **tissue culture**. This is used, for example, to propagate orchids, which are very difficult to grow from seeds. A small number of cells is taken from the plant, and put onto a sterile jelly that contains nutrients and plant hormones (Section 9.25). The hormones stimulate the cells to divide by mitosis and form a shapeless lump called a **callus**. Then the callus is moved onto a different jelly, containing different hormones, which stimulate it to grow roots. It may then be moved to another jelly, where it grows shoots. In this way, very large numbers of plants can be grown from a single plant.

Sexual reproduction

8.8 Sexual reproduction involves fertilisation.

In sexual reproduction, each parent organism produces sex cells called **gametes** (Fig 8.7). Eggs and sperm are examples of gametes. Two of these gametes then join together. This is called **fertilisation**. The new cell which is formed by fertilisation is called a **zygote**. The zygote divides again and again, and eventually grows into a new organism.

The new organisms produced by sexual reproduction are not identical to each other or to their parents. They contain different combinations of genes. You will see why this is so in Section 8.10.

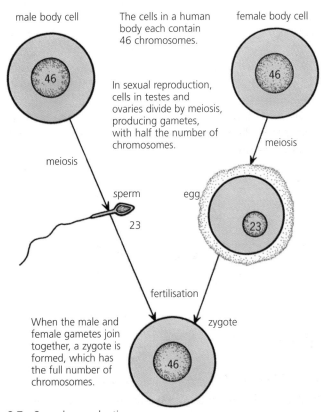

8.7 Sexual reproduction.

8.9 Gametes have half the normal number of chromosomes.

Gametes are different from ordinary cells, because they contain only half as many chromosomes as usual. This means that when two of them fuse together, the zygote they form will have the correct number of chromosomes.

Humans, for example, have 46 chromosomes in each of their body cells. But human egg and sperm cells only have 23 chromosomes each. When an egg and sperm fuse together at fertilisation, the zygote which is formed will therefore have 46 chromosomes, the normal number (Fig 8.7).

The 46 chromosomes in an ordinary human cell are of 23 different kinds. There are two of each kind. A cell which has the full number of chromosomes, with two complete sets, is called a **diploid cell**.

An egg or sperm, though, only has 23 chromosomes, one of each kind. A cell like this, with a single set of chromosomes, is called a **haploid cell**. Gametes are always haploid. When two gametes fuse together, they form a **diploid zygote**.

8.10 Gametes are made by meiosis.

Gametes are made by ordinary body cells dividing. For example, human sperm are made when cells in a testis divide.

Because gametes need to have only half as many chromosomes as their parent cell, division by mitosis will not do. When gametes are being made, cells divide in a different way, called **meiosis**. Meiosis is sometimes called **reduction division**, because it halves (reduces) the number of chromosomes in a cell. This process is shown in Fig 8.8.

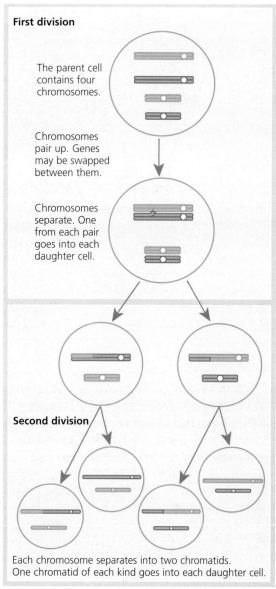

8.8 Summary of chromosome behaviour during meiosis.

In flowering plants and animals, meiosis only happens when gametes are being made. Meiosis produces new cells with only half as many chromosomes as the parent cell.

If you look carefully at Fig 8.8, you will see that the pairs of chromosomes have swapped some pieces between themselves. These pieces may carry different versions of genes. This means that the gametes which are produced carry different combinations of genes – they do not have the same combination as the cell which made them, nor the same as each other. You can find out more about how this can produce variation amongst the offspring produced by sexual reproduction in Chapter 12.

8.11 Male gametes move – female ones stay still.

In many organisms, there are two different kinds of gamete. One kind is quite large, and does not move much. This is called the female gamete. In humans, the female gamete is the egg.

The other sort of gamete is smaller, and moves actively in search of the female gamete. This is called the male gamete. In humans, the male gamete is the sperm.

Often, each individual organism can only produce one kind of gamete. Its gender is either male or female, depending on what kind of gamete it makes. All mammals, for example, are either male or female.

Sometimes, though, a organism can produce both sorts of gamete. Earthworms, for example, can produce both eggs and sperm. An organism which produces both male and female gametes is a **hermaphrodite**. Many flowering plants are also hermaphrodite.

Sexual reproduction in a mammal

8.12 The female reproductive organs.

Fig 8.9 shows the reproductive organs of a woman. The female gametes, called **eggs**, are made in the two **ovaries**. Leading away from the ovaries are the **oviducts**, sometimes called Fallopian tubes. They do not connect directly to the ovaries, but have a funnel-shaped opening just a short distance away.

The two oviducts lead to the womb or **uterus**. This has very thick walls, made of muscle. It is quite small – only about the size of a clenched fist – but it can stretch a great deal when a woman is pregnant.

At the base of the uterus is a narrow opening, guarded by muscles. This is the neck of the womb, or **cervix**. It leads to the **vagina**, which opens to the outside.

The opening from the bladder, called the **urethra**, runs in front of the vagina, while the **rectum** is just behind it. The three tubes open quite separately to the outside.

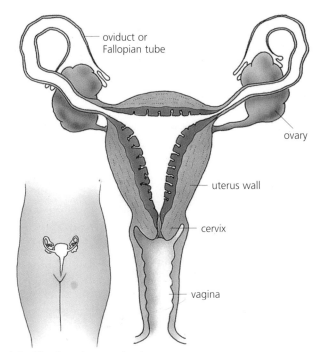

8.9 The female reproductive organs.

8.13 The male reproductive organs.

Fig 8.10 shows the reproductive organs of a man. The male gametes, called **spermatozoa** or **sperm**, are made in two **testes**. These are outside the body, in two sacs of skin called the **scrotum**.

The sperm are carried away from each testis in a tube called the **sperm duct**. The sperm ducts from the testes join up with the **urethra** just below the bladder. The urethra continues downwards, and opens at the tip of the **penis**. The urethra can carry both urine and sperm at different times.

Where the sperm ducts join the urethra, there is a gland called the **prostate gland**. This makes a fluid which the sperm swim in. Just behind the prostate gland are the **seminal vesicles**, which also secrete fluid.

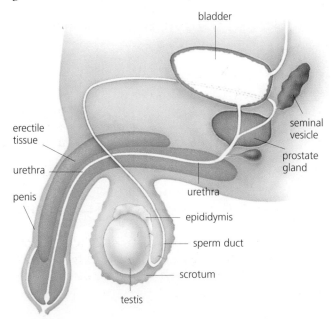

8.10 The male reproductive organs.

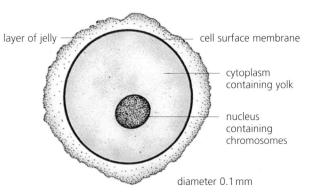

8.11 An egg.

8.14 Ovaries make eggs.

Eggs begin to be formed inside a girl's ovaries before she is born. At birth, she will already have many thousands of partly-developed eggs inside her ovaries.

When she reaches puberty, some of these eggs will begin to develop. Usually, only one develops at a time. When it is mature (Fig 8.11), an egg bursts out of the ovary and into the funnel at the end of the oviduct. Usually only one egg, from one ovary, does this at a time. This is called **ovulation**. In humans, it happens about once a month.

8.15 Testes make sperm.

Inside the testes, very large numbers of sperm cells are formed every day. Sperm are made continually from puberty onwards. Fig 8.12 shows the structure of a sperm.

Sperm production is very sensitive to heat. If they get too hot, the cells in the tubules within the testis will not develop into sperm. This is why the testes are outside the body, where they are cooler than they would be inside.

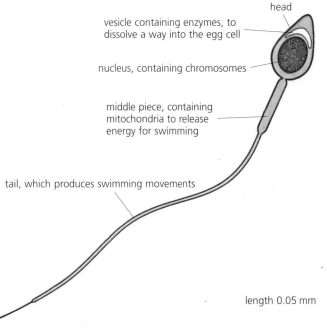

8.12 A sperm cell.

8.16 Mating introduces sperm into the vagina.

After ovulation, the egg is caught in the funnel of the oviduct. The funnel is lined with cilia which beat rhythmically, wafting the egg into the entrance of the oviduct. Very slowly, the egg travels towards the uterus. Cilia lining the oviduct help to sweep it along (Fig 8.13). Muscles in the wall of the oviduct also help to move it, by peristalsis (Section 4.33).

If the egg is not fertilised by a sperm within 8–24 hours after ovulation, it will die. By this time, it has only travelled a short way along the oviduct. So a sperm must reach an egg while it is quite near the top of the oviduct if fertilisation is to be successful.

When the man is sexually excited, blood is pumped into spaces inside the penis, so that it becomes erect. To bring the sperm as close as possible to the egg, the man's penis is placed inside the vagina of the woman.

Sperm are pushed out of the penis into the vagina. This happens when muscles in the walls of the tubes containing the sperm contract rhythmically. The wave of contraction begins in the testes, travels along the sperm ducts, and into the penis. The sperm are squeezed along, out of the man's urethra and into the woman's vagina. This is called **ejaculation**.

The fluid containing the sperm is called **semen**. Ejaculation deposits the semen at the top of the vagina, near the cervix.

Questions
1 What is the name for the narrow opening between the uterus and the vagina?
2 Where is the prostate gland, and what is its function?

8.17 Fertilisation happens in the oviduct.

The sperm are still quite a long way from the egg. They swim, using their tails, up through the cervix, through the uterus, and into the oviduct (Fig 8.14).

Sperm can only swim at a rate of about 4 mm per minute, so it takes quite a while for them to get as far as the oviducts. Many will never get there at all. But one ejaculation deposits about a million sperm in the vagina, so there is a good chance that some of them will reach the egg.

One sperm enters the egg. The nucleus of the sperm fuses with the nucleus of the egg. This is **fertilisation** (Fig 8.15).

As soon as the successful sperm enters the egg, the egg membrane becomes impenetrable, so that no other sperm can get in. The unsuccessful sperm will all die.

8.13 Scanning electron micrograph of sperm cells swimming over the ciliated cells of the oviduct.

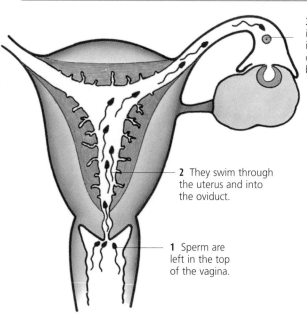

8.14 How sperm get to the egg (sperm and egg drawn to different scale).

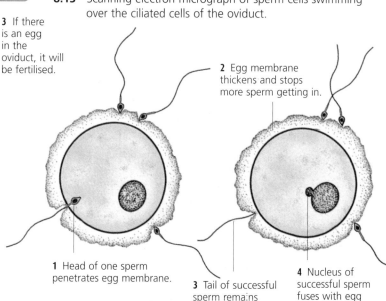

8.15 Fertilisation.

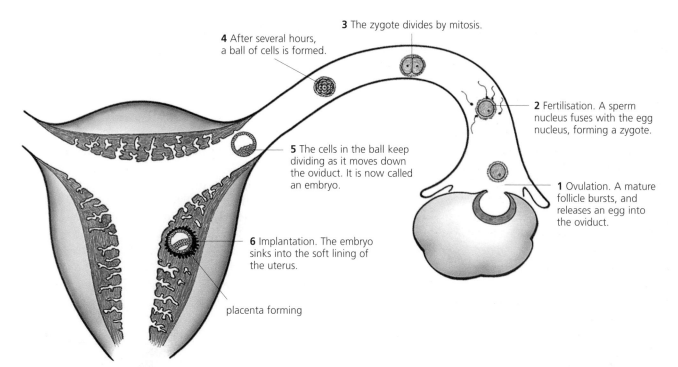

8.16 Stages leading to implantation.

8.18 The zygote implants in the uterus wall.

When the sperm nucleus and the egg nucleus have fused together, they form a zygote. The zygote continues to move slowly down the oviduct. As it goes, it divides by mitosis. After several hours, it has formed a ball of cells. This is called an **embryo**. The embryo obtains food from the yolk of the egg.

It takes several hours for the embryo to reach the uterus, and by this time it is a ball of 16 or 32 cells. The uterus has a thick, spongy lining, and the embryo sinks into it. This is called **implantation** (Fig 8.16).

8.19 The fetus's life-support system is its placenta.

The cells in the embryo, now buried in the soft lining of the uterus, continue to divide. As the embryo grows, a **placenta** also grows, which connects it to the wall of the uterus (Figs 8.17 and 8.18). The placenta is soft and dark red, and has finger-like projections called **villi**. The villi fit closely into the uterus wall.

After eleven weeks, the embryo has grown into a **fetus**. The placenta is joined to the fetus by the **umbilical cord**. Inside the cord are two arteries and a vein. The arteries take blood from the fetus into the placenta, and the vein returns the blood to the fetus.

In the placenta are capillaries filled with the fetus's blood. In the wall of the uterus are large spaces filled with the mother's blood. The fetus's and mother's blood do not mix. They are separated by the wall of the placenta. But they are brought very close together, because the wall of the placenta is very thin.

Questions

1. Explain how ovulation happens.
2. Where are sperm made?
3. How does an egg travel along the oviduct?
4. What is semen?
5. Where does fertilisation take place?

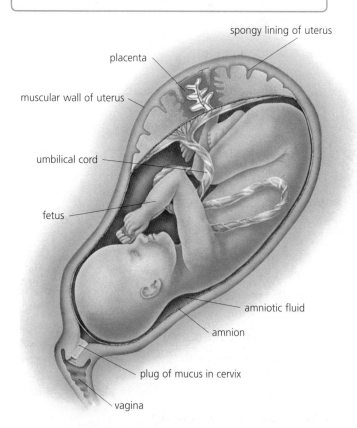

8.17 Side view of developing fetus inside uterus.

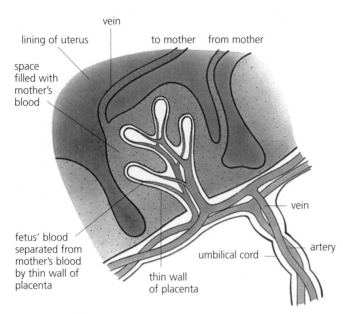

8.18 Part of the placenta.

Oxygen and soluble nutrients, such as glucose, amino acids, vitamins and mineral ions in the mother's blood diffuse across the placenta into the fetus's blood, and are then carried along the umbilical cord to the fetus. Carbon dioxide and urea diffuse the other way, and are carried away in the mother's blood.

As the fetus grows, the placenta grows too. By the time the baby is born, the placenta will be a flat disc, about 12 cm in diameter, and 3 cm thick.

8.20 An amnion protects the fetus.

The fetus is surrounded by a strong membrane, called the **amnion**. The amnion makes a liquid called **amniotic fluid**. This fluid helps to support the fetus, and to protect it.

8.21 A baby develops during gestation.

No-one fully understands how the cells in the ball which embedded itself in the wall of the uterus become arranged to form a baby. The cells gradually divide and grow. By eleven weeks after fertilisation they have become organised into all the different organs.

After this, the fetus just grows. It takes nine months before it is ready to be born. This length of time between fertilisation and birth is called the **gestation period**.

8.22 Muscular contractions cause birth.

A few weeks before birth, the fetus usually turns over in the uterus, so that it is lying head downwards. Its head lies just over the opening of the cervix. Birth begins when the strong muscles in the wall of the uterus start to contract. This is called **labour**.

To begin with, the contractions are quite gentle, and may only happen about once an hour. Gradually, they become stronger and more frequent. The contractions of the muscles slowly stretch the opening of the cervix (Fig 8.19).

After several hours, the cervix is wide enough for the head of the baby to pass through. Now, the muscles start to push the baby down through the cervix and the vagina. This part of the birth happens quite quickly.

The baby is still attached to the uterus by the umbilical cord and the placenta. Now that it is in the open air, it can breathe for itself, so the placenta is no longer needed. The placenta falls away from the wall of the uterus, and passes out through the vagina. It is called the **afterbirth**.

The umbilical cord is cut, and clamped just above the point where it joins the baby. This is completely painless, because there are no nerves in the cord. The stump of the cord forms the baby's navel.

The contractions of the muscles of the uterus are painful. They feel rather like cramp. The mother can help herself a lot by preparing her body with exercises before labour begins and by breathing in a special way during labour, and she can also be given pain-killing drugs if she needs them.

> **Questions**
> 1 What is formed when an egg and sperm fuse together?
> 2 What kind of cell division takes place in the growth of an embryo?
> 3 From where does the very young embryo obtain its food?

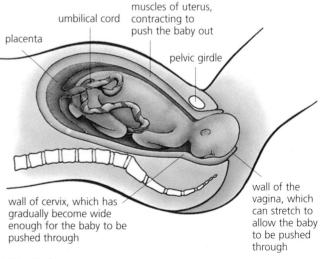

8.19 Birth.

8.23 Pregnant women should care for their health.

When a woman is pregnant, she should take extra care of her health, both for her own benefit and that of her baby. This is sometimes called **ante-natal care**, meaning 'before birth'.

She should ensure that her diet contains plenty of calcium, to help to form the growing fetus's bones. She also needs extra iron, because her body will produce a lot of extra blood to help to carry oxygen and nutrients to the placenta, and her growing baby is also forming blood. Iron is needed to make the haemoglobin in the red blood cells. She may also need a little extra carbohydrate, because she needs extra energy to help to move her heavier body around, and extra protein, to help to form her growing fetus's new cells.

She should continue to take exercise. Most people consider that steady, gentle exercise is best, such as swimming or walking. She may also be given special exercises to do which will help her to stay fit during pregnancy, and also allow her to take an active part when she is giving birth.

We have seen that many useful substances cross the placenta from the mother's blood to the fetus's blood. Unfortunately, harmful substances can cross, too. For example, if the mother smokes, nicotine and carbon monoxide can enter the baby's blood, and this can cause the baby to grow more slowly and be born smaller than if the mother was a non-smoker. A woman should never smoke during pregnancy. She also needs to take care not to drink too much alcohol, or to take any drug without advice from her doctor.

The mother also needs to avoid some illnesses. Rubella is caused by a virus, producing a rash and a fever. If the rubella virus crosses the placenta, it can cause serious harm to the fetus, who may be born deaf or with other disabilities. In many countries, teenage girls are offered vaccination against rubella. Another disease which is very dangerous to a fetus is AIDS. The virus that causes AIDS, called HIV, can cross the placenta so that a baby may be born with AIDS itself.

8.24 Mammals care for their young.

Although it has been developing for nine months, a human baby is very helpless when it is born. Usually, both parents help to care for it.

During pregnancy, the glands in the mother's breasts will have become larger. Soon after the birth of the baby, they begin to make milk. This is called **lactation**. Lactation happens in all mammals, but not in other animals (Fig 8.20).

Most people consider that feeding a baby on breast milk is much better than feeding it on bottle milk. Bottle milk, sometimes called formula, is bought as powder that is mixed with water and sterilised. The baby then sucks this milk from a bottle.

Although this can make life easier for the mother, because she can hand over the feeding of her baby to someone else, it does have several disadvantages. For example, it is much more expensive than breast milk, which is free! And, unless the equipment used for making up the bottle milk is kept clean, it is easy for bacteria to get into the milk and make the baby ill.

Another advantage of breast milk is that it contains antibodies from the mother, which help the baby to fight off infectious diseases. (Antibodies are described in Section 13.10.) Breast feeding also helps a close relationship to develop between the mother and her baby, which is beneficial to both of them.

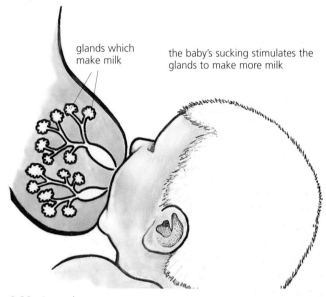

8.20 Lactation.

Questions

1. What is (a) implantation, (b) a fetus?
2. How is the fetus connected to the placenta?
3. Describe two ways in which the structure of the placenta helps diffusion between the mother's and the fetus's blood to take place quickly.
4. List two substances which pass from the mother's blood into the fetus's blood.
5. What is the function of the amnion?
6. How long is the gestation period in humans?
7. Describe what happens to each of the following during the birth of a baby.
    (a) muscles in the uterus wall
    (b) the cervix
    (c) the placenta
8. Draw a comparison table to show the advantages of breast feeding compared with bottle feeding.

8.25 The menstrual cycle lasts about 28 days.

Usually, one egg is released into the oviduct every month in an adult woman (Fig 8.21). Before the egg is released, the lining of the uterus becomes thick and spongy, to prepare itself for the fertilised egg. It is full of tiny blood vessels, ready to supply the embryo with food and oxygen if it should arrive.

If the egg is not fertilised, it is dead by the time it reaches the uterus. It does not sink into the spongy wall, but continues onwards, down through the vagina. As the spongy lining is not needed now, it gradually disintegrates. It, too, is slowly lost through the vagina. This is called **menstruation**, or a period. It usually lasts for about five days.

After menstruation, the lining of the uterus builds up again, so that it will be ready to receive the next egg, if it is fertilised. In many women the menstrual cycle lasts about 28 days. However, it is common for women to have cycles that are shorter or longer than this. And, while some women have very regular and predictable menstrual cycles, in others the cycles are much less regular.

The menstrual cycle is controlled by hormones. This is described in Section 9.21.

8.26 Sexual maturity is reached at puberty.

The time when a person approaches sexual maturity is called **adolescence**. Sperm production begins in a boy, and ovulation in a girl.

During adolescence, the secondary sexual characteristics develop. In boys, these include growth of facial and pubic hair, breaking of the voice, and muscular development. In girls, pubic hair begins to grow, the breasts develop, and the pelvic girdle becomes broader.

These changes are brought about by hormones. The male hormones are called **androgens** of which the most important is **testosterone**. They are produced in the testes. The female hormones are called **oestrogens**. They are produced in the ovaries.

The point at which sexual maturity is reached is called **puberty**. This is often several years earlier for girls than for boys. At puberty, a person is still not completely adult, because emotional development is not complete.

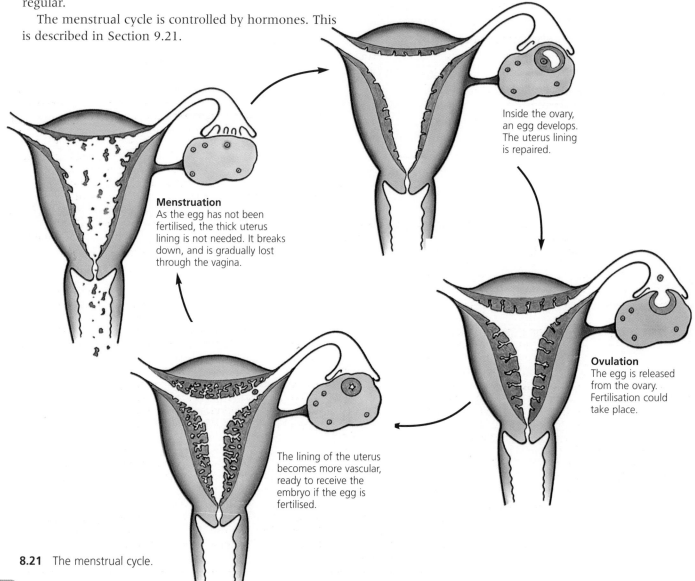

Menstruation
As the egg has not been fertilised, the thick uterus lining is not needed. It breaks down, and is gradually lost through the vagina.

Inside the ovary, an egg develops. The uterus lining is repaired.

Ovulation
The egg is released from the ovary. Fertilisation could take place.

The lining of the uterus becomes more vascular, ready to receive the embryo if the egg is fertilised.

8.21 The menstrual cycle.

8.27 Some diseases are transmitted by sexual intercourse.

The viruses or bacteria that cause some infectious diseases can be passed from one person to another during sexual intercourse. The diseases they cause are called **sexually transmitted diseases**. They include gonorrhoea, syphilis and AIDS. These are described in Sections 13.6 and 13.14.

8.28 Birth control prevents unwanted pregnancies.

Birth control can help couples to have no more children than they want. Birth control is important in keeping family sizes small, and in limiting the increase in the human population. Careful and responsible use of birth control methods means no unwanted children are born.

- **Mechanical methods** work by putting a barrier between the eggs and sperm. Two examples are the condom and the cap.
- **Hormonal methods** use hormones to stop eggs being produced. These hormones are like those that a woman's body makes when she is pregnant; that is, progesterone and oestrogen. The woman takes a pill each day, containing these hormones. A woman on the pill has no eggs present to be fertilised.
- **Surgical methods** tend to be most suitable for couples who already have as many children as they want. The operation for a man is called a vasectomy. It is a quick and simple operation, usually done under local anaesthetic. The operation for a woman usually involves a short stay in hospital, and a general anaesthetic.
- **Chemical methods** use chemicals called spermicides, which kill sperm. They are best used in combination with another method. For example, spermicides may be inserted into the vagina with a cap.
- **Natural methods** involve the couple ensuring that they do not have sexual intercourse when the woman has an egg in her oviducts. It is a risky method, and only works for women who have very regular and predictable menstrual cycles. However, it is useful for couples who do not wish to use other birth control methods for religious or other reasons.

> **Questions**
> 1 What is meant by (a) adolescence (b) puberty?
> 2 What are androgens?
> 3 List two effects of androgens.

Table 8.1 Some methods of birth control.

Method	How it works	Advantages and disadvantages
Condom (mechanical)	The condom is placed over the erect penis. It traps semen as it is released, stopping it from entering the vagina.	This is a very safe method of contraception if used correctly, but care must be taken that no semen is allowed to escape before it is put on or after it is removed. It can also help to prevent the transfer of infection, such as gonorrhoea and HIV, from one partner to the other.
Cap (mechanical)	The cap is a circular sheet of rubber, which is placed over the cervix, at the top of the vagina. Spermicidal (sperm-killing) cream is first applied round its edges. Sperm deposited in the vagina cannot get past the cap into the uterus.	This is an effective method, if used and fitted correctly. Fitting must be done by a doctor, but after that a woman can put her own cap in and take it out as needed.
The pill or oral contraceptive (hormonal)	The pill contains the female sex hormones oestrogen and progesterone. One pill is taken every day. The hormones are like those that are made when a woman is pregnant, and stop egg production.	This is a very effective method, so long as the pills are taken at the right time. However, some women do experience unpleasant side effects, and it is important that women on the pill have regular check-ups with their doctor.
Sterilisation (surgical)	In a man, the sperm ducts are cut or tied, stopping sperm from travelling from the testes to the penis. In a woman, the oviducts are cut or tied, stopping eggs from travelling down the oviducts.	An extremely sure method of contraception, with no side effects. However, the tubes often cannot be re-opened if the person later decides they do want to have children, so it is not a method for young people.
Spermicides (chemical)	Spermicidal cream in the vagina kills sperm.	Quite easy to use. Only effective, however, if used in combination with another method, such as the cap.
Natural	The woman keeps a careful record of her menstrual cycle over several months, so that she can predict roughly when an egg is likely to be present in her oviducts. She must avoid sexual intercourse for several days around this time.	This is a very unsafe method, because it is never possible to be 100% certain when ovulation is going to happen. Nevertheless, it is used by many people who do not want to use one of the other contraceptive methods.

8.29 Hormones can be used to increase fertility.

Whereas many couples want to use birth control methods to limit the number of children that they have, others have the opposite problem – they are not able to have children. The problem that is causing the couple's infertility may be in the man or in the woman. For example, the man may not be producing healthy sperm. If this is the case, then the couple may decide that they will have a baby using sperm from another man. Sperm from a donor is collected in a clinic, and can be stored at a low temperature for many months or even years. The woman can then attend the clinic, and some of the sperm can be placed into her vagina. This is called **artificial insemination**.

This may be a real help to a couple, as it allows them to have a child that they could not otherwise have. However, they need to think very carefully about this before they go ahead, and make sure that they are both happy with the idea. The man has to be able to accept that the child they have is not his. Problems can also be caused when the child grows up and wants to know who his or her father is. It can be very difficult for a young person not to know this, so some people think that the identity of the sperm donor should be given to the child. Others, however, think this may cause more problems than it solves, because one sperm donor could end up being the father of many children. Indeed, fewer people would be likely to become sperm donors if this information was freely available.

Another way in which an infertile couple can be helped is using **fertility drugs**. This method is used when the woman is not producing enough eggs. She is given hormones, including one called FSH, that cause her to produce eggs. Sometimes, these are simply allowed to be released into the oviducts in the normal way. Sometimes, they are removed from her ovaries just before they are due to be released, and placed in a warm liquid in a petri dish. Some of her partner's sperm are then added, and fertilisation takes place in the dish. Two or three of the resulting zygotes are then placed into her uterus, where they develop in the usual way.

This method is quite expensive, and some people think that it should not be freely available to anyone who wants it. Others think that the inability to have children can be so devastating to a couple that they should receive the treatment free of charge. The treatment is not always successful, and may have to be repeated many times before a woman becomes pregnant. Another problem is that while usually only one of the embryos develops, sometimes two or three do, so that the couple has twins or triplets when they really only wanted one child.

Sexual reproduction in flowering plants

8.30 Flowers are for sexual reproduction.

Many flowering plants can reproduce in more than one way. Often, they can reproduce asexually and also sexually by means of flowers.

The function of a flower is to make gametes, and to ensure that fertilisation will take place. Fig 8.22 illustrates the structure of an insect-pollinated flower. On the outside of the flower are four **sepals**. The sepals protect the flower while it is a bud.

Just inside the sepals are four **petals**. These are brightly coloured, and have lines on them running from top to bottom. The petals attract insects to the flower. The lines are called **guide-lines**, because they guide the insect to the base of the petal. Here, there is a gland called a **nectary**. The nectary makes a sugary liquid called nectar, which insects feed on.

Inside the petals are six **stamens**. These are the male parts of the flower. Each stamen is made up of a long **filament**, with an **anther** at the top. The anthers contain **pollen grains**, which contain the male gametes.

The female part of the flower is in the centre. It is called a **carpel**. Most of it consists of an **ovary**. Inside the ovary are many **ovules**, which contain the female gametes (Fig 8.24). At the top of the ovary is a short **style**, with a forked **stigma** at the tip. The function of the stigma is to catch pollen grains.

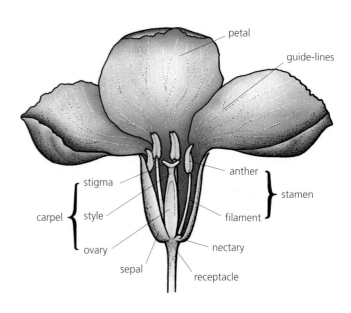

8.22 A flower with one petal removed.

8.31 Pollen grains contain male gametes.

The male gametes are inside the pollen grains, which are made in the anthers.

Figs 8.23a and b illustrate a young anther, as it looks before the flower bud opens. The anther has four spaces or **pollen sacs** inside it. Some of the cells around the edge of the pollen sacs divide by meiosis to make pollen grains. When the flower bud opens, the anthers split open (Fig 8.23c). Now the pollen is on the outside of the anther.

Pollen looks like a fine, yellow powder. Under the microscope, you can see the shape of individual grains. Pollen grains from different kinds of flowers have different shapes.

Each grain is surrounded by a hard coat, so that it can survive in difficult conditions if necessary. The pollen of this flower has a sticky coat, so that it will stick to insects' bodies.

8.32 Each ovule contains a female gamete.

The female gametes are inside the ovules, in the ovary. They have been made by meiosis. Each ovule contains just one gamete (Fig 8.24).

8.33 Pollen must be carried from anther to stigma.

For fertilisation to take place, the male gametes must travel to the female gametes. The first stage of this journey is for pollen to be taken from the anther where it was made, to a stigma. This is called **pollination**.

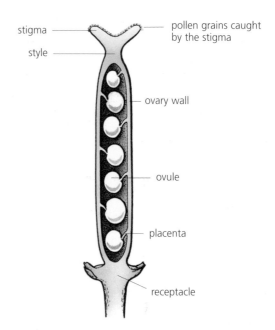

8.24 Section through the carpel of the flower in Fig 8.22.

In many flowers, pollination is carried out by insects. Small insects, such as beetles and honey bees, come to the flowers, attracted by their colour and strong, sweet scent. The bee follows the guide-lines to the nectaries, brushing past the anthers as it goes. Some of the pollen will stick to its body.

The bee will probably then go to another flower, looking for more nectar. Some of the pollen it picked up at the first flower will stick onto the stigma of the second flower when the bee brushes past it. The stigma is sticky, and many pollen grains get stuck on it (Fig 8.25).

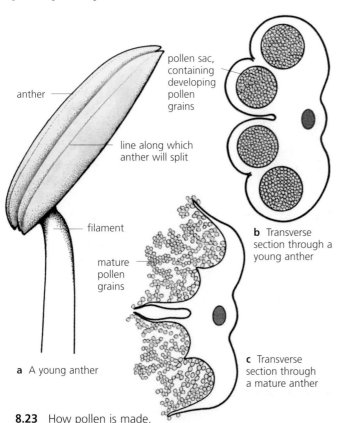

8.23 How pollen is made.

8.25 Scanning electron micrograph of pollen on a stigma.

8.34 Flowers can be self- or cross-pollinated.

Sometimes, pollen is carried to the stigma of the same flower, or to another flower on the same plant. This is called **self-pollination**.

If pollen is taken to a flower on a different plant of the same species, this is called **cross-pollination**. If pollen lands on the stigma of a different species of plant, it usually dies.

Self-pollination produces less variation amongst the resulting offspring than cross-pollination. You can read about the importance of variation in natural selection and evolution in Chapter 12.

8.35 Some flowers are wind-pollinated.

In an insect-pollinated flower, pollen is carried from an anther to a stigma by insects. In some flowers, it is the wind which does this.

Fig 8.26 illustrates a grass flower, which is an example of a wind-pollinated flower. Table 8.2 (overleaf) compares insect-pollinated and wind-pollinated flowers.

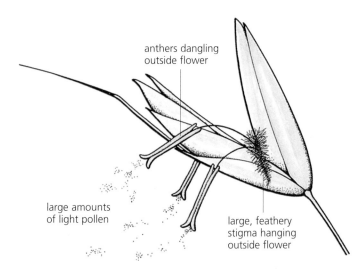

8.26 An example of a wind-pollinated flower – a grass flower.

8.36 Pollen tubes take male gametes to ovules.

After pollination, the male gamete inside the pollen grain on the stigma still has not reached the female gamete. The female gamete is inside the ovule, and the ovule is inside the ovary.

If it has landed on the right kind of stigma, the pollen grain begins to grow a tube. You can try growing some pollen tubes, in Practical 8.2. The pollen tube grows down through the style and the ovary, towards the ovule (Fig 8.27). It secretes enzymes to digest a pathway through the style.

The ovule is surrounded by a double layer of cells called the **integuments**. At one end, there is a small hole in the integuments, called the **micropyle**. The pollen tube grows through the micropyle, into the ovule.

The male gamete travels along the pollen tube, and into the ovule. It fuses with the female gamete. Fertilisation has now taken place.

One pollen grain can only fertilise one ovule. If there are many ovules in the ovary, then many pollen grains will be needed to fertilise them all.

Practical 8.1 Investigating the structure of an insect-pollinated flower

1. Take an open, fresh looking insect-pollinated flower. Can you suggest two ways in which the flower advertises itself to insects?
2. Gently remove the sepals from the outside of the flower. Look at the sepals on a flower bud. What is the function of the sepals?
3. Now remove the petals from your flower. Make a labelled drawing of one of them, to show the markings. What is the function of these markings?
4. Find the stamens. If you have a young flower there will be pollen on the anthers at the top of the stamens. Dust some onto a microscope slide, and look at it under a microscope. Draw a few pollen grains.
5. Now remove the stamens. What do you think is the function of the filaments?
6. Using a hand lens, look for the nectaries at the bottom of the flower. What is their function?
7. The carpel is now all that is left of the flower. Find an ovary, style and stigma. Look at the stigma under a binocular microscope. What is its function, and how is it adapted to perform it?
8. Using a sharp razor blade, make a clean cut lengthways through the ovary, style and stigma. You have made a longitudinal section. Find the ovules inside the ovary. How big are they? What colour are they? About how many are there?

Practical 8.2 Growing pollen tubes

When a stigma is ripe, it secretes a fluid which stimulates pollen grains on it to grow tubes. The fluid contains sugar. In this experiment, you can try germinating different kinds of pollen grains in different concentrations of sugar solution.

It is best if the class is divided into groups. Each group should use sugar solution of just one concentration.

	Concentration of sugar solution / mol dm^{-3}				
	distilled water	0.1	0.5	1.0	etc
Flower A					
Flower B					
Flower C					
Flower D					

1. Collect four cavity slides. Using your finger, make a neat ring of petroleum jelly around the outer edge of each cavity.
2. Stick a label on each slide. Write your initials on it, and the concentration of sugar solution your group is using.
3. Fill the cavity in each slide with sugar solution.
4. Choose one flower of each kind which has pollen on its anthers. Dust pollen from one flower onto the solution on one of your slides. Gently lower a coverslip over it, without squashing the petroleum jelly ring. Write the name of the flower on the label.
5. Repeat step 4 with the other three flowers.
6. Place each slide in a warm incubator, and leave for at least an hour.
7. Set up a microscope. Examine each of your slides under the microscope. Look carefully for pollen tubes. Record your results in the table, and collect results from groups using other concentrations of sugar solution.

Questions

1. Why was a ring of petroleum jelly put around the cavity in each slide?
2. In which solution did each of the four types of pollen germinate best?
3. Can you suggest why pollen dies if it lands on an unripe stigma, or a stigma of the wrong sort of flower?
4. Why do pollen grains grow tubes?

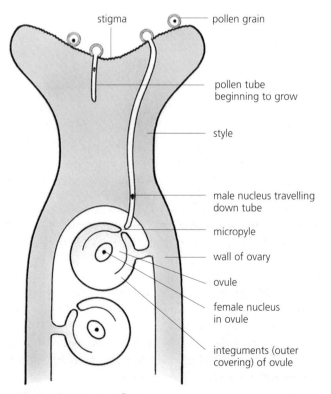

8.27 Fertilisation in a flower.

Questions

1. What is the function of a flower?
2. In which part of a flower are male gametes made?
3. In which part of a flower are female gametes made?
4. What is pollination?
5. Why do wind-pollinated flowers usually produce more pollen than insect-pollinated ones?
6. After pollination, how does the male gamete reach the ovule?
7. What is a micropyle?

Table 8.2 A comparison between wind-pollinated and insect-pollinated flowers.

Insect-pollinated	Wind-pollinated
Large, conspicuous petals, often with guide-lines	Small, inconspicuous petals, or no petals at all
Often strongly scented	No scent
Often have nectaries at base of flowers	No nectaries
Anthers inside flower, where insect has to brush past them to get nectar	Anthers dangling outside flower, where they catch the wind
Stigma inside flower, where insect has to brush past it to get to nectar	Stigma large and feathery, dangling outside flower, where pollen in the air may land on it
Sticky or spiky pollen grains, which will stick to insects	Smooth, light pollen, which can be blown in the wind
Quite large quantities of pollen made, because some will be eaten, or carried to the wrong sort of flower	Very large quantities of pollen made, because most will be blown away and lost
Flowers usually appear at warm times of year, when there are plenty of active insects	Flowers sometimes appear at colder times of year

8.37 Fertilised ovules become seeds.

Once the ovules have been fertilised, many of the parts of the flower are not needed any more. The sepals, petals and stamens have all done their job. They wither, and fall off.

Inside the ovary, the ovules start to grow. Each ovule now contains a zygote, which was formed at fertilisation. The zygote divides by mitosis to form an embryo plant.

The ovule is now called a **seed**. The integuments of the ovule become hard and dry, to form the testa of the seed. Water is withdrawn from the seed, so that it becomes dormant (Section 8.41).

The ovary also grows. It is now called a **fruit**. The wall of the fruit is called the **pericarp**.

8.38 Fruits protect and disperse seeds.

The function of the fruit is to protect the seeds inside it until they are ripe, and then to help disperse the seeds (Fig 8.28). Dispersal of seeds is important, because it prevents too many plants growing close together. If this happens, they compete for light, water and nutrients, so that none of them can grow properly. Dispersal also allows the plant to colonise new areas.

8.39 Fruits are ovaries after fertilisation.

Plants have an enormous variety of fruits, all adapted to disperse their seeds as effectively as possible.

The biological definition of a fruit is an ovary after fertilisation, containing seeds. Mangoes, plums and oranges are true fruits, but so also are tomatoes, cucumbers and pea pods! You can tell a fruit because (a) it contains one or more seeds, and (b) it has two scars – one where it was attached to the plant, and one where the style and stigma were attached to it.

Sometimes, it is not easy to tell a fruit from a seed. A seed, though, only has one scar, called the **hilum**, where it was joined onto the fruit.

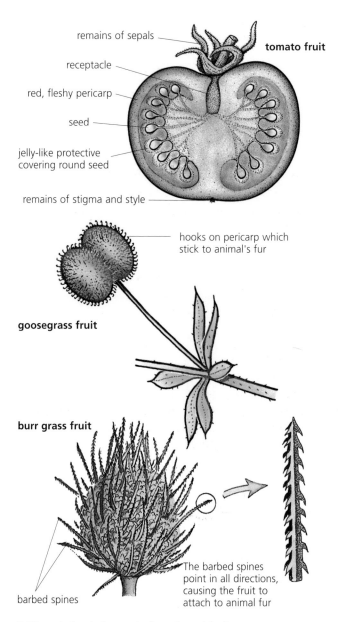

8.28a Animal dispersal of seeds and fruits.

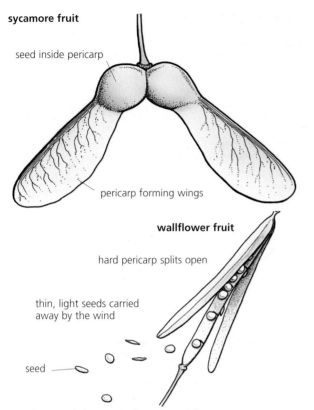

8.28b Wind dispersal of seeds and fruits.

8.40 A seed contains an embryo plant.

Figure 8.29 shows the structure of a French bean seed. The seed has developed from a fertilised ovule.

A seed contains an embryo plant. The embryo consists of a **radicle**, which will grow into a root, and a **plumule**, which will grow into a shoot.

There is also food for the embryo. In a French bean seed, the food is stored in two cream-coloured **cotyledons**. These contain starch and protein. The cotyledons also contain enzymes.

Surrounding the cotyledons is a tough, protective covering called the **testa.** The testa stops the embryo from being damaged and it prevents bacteria and fungi from entering the seed.

The testa has a tiny hole in it called the **micropyle**. Near the micropyle is a scar, the **hilum**, where the seed was joined onto the pod.

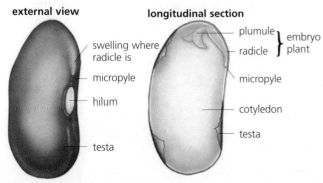

8.29 Structure of a French bean seed.

Questions
1. Give two functions of a fruit.
2. List two different ways in which seeds may be dispersed, giving one example for each, and explaining how the fruit is adapted for this method of dispersal.
3. Give two differences between fruits and seeds.
4. Which are fruits, and which are not?
   (a) apple (e) lemon
   (b) tomato (f) bean pod
   (c) potato (g) cucumber
   (d) cabbage

8.41 Uptake of water begins seed germination.

A seed contains hardly any water. When it was formed on the plant, the water in it was drawn out, so that it became dehydrated. Without water, almost no metabolic reactions can go on inside it. The seed is inactive or **dormant**. This is very useful, because it means that the seed can survive harsh conditions, such as cold or drought, which would kill a growing plant.

A seed must have certain conditions before it will begin to germinate. You can find out what they are if you do Practical 8.3.

When a seed germinates, it first takes up water through the micropyle. As the water goes into the cotyledons, they swell. Eventually, they burst the testa (Fig 8.31).

8.42 During germination, enzymes digest food stores.

When a seed first begins to germinate, it increases in mass. This is because it absorbs water from the soil.

As soon as it begins to grow, it starts to use its food stores. Once there is sufficient water, the enzymes in the cotyledons become active. Amylase begins to break down the stored starch molecules to maltose. Proteases break down the protein molecules to amino acids.

Maltose and amino acids are soluble, so they dissolve in the water. They diffuse to the embryo plant, which uses these foods for growth. The amino acids are used to make new protein molecules for cell membranes and cytoplasm. Some of the stored sugar is made into cellulose, to make cell walls for new cells.

All this requires energy. The seed, like all living organisms, gets its energy by breaking down glucose, in respiration. Quite a lot of the glucose from the stored starch is used up in respiration, so the seed loses weight.

After a few days, the plumule of the seed grows above the surface of the ground. The first leaves open out and begin to photosynthesise. The plant can now make its own food faster than it is using it up. It begins to increase in mass.

Fig 8.30 summarises the changes in mass of an annual plant, such as a French bean, from germination until death. An annual plant is one which lives for less than one year.

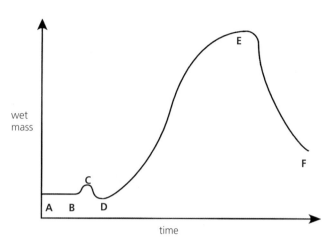

A to **B**: seed is dormant
B to **C**: germination begins, and seed increases in mass as it absorbs water
C to **D**: seed loses mass, as it uses up food stores to provide energy for growth
D to **E**: plant is photosynthesising, building up new cells
E to **F**: plant loses mass quickly, as it flowers and produces seeds and fruits, which are dispersed; plant gradually dies

8.30 Growth curve for an annual plant.

8.43 Growth is an increase in dry mass.

When an organism grows, its cells divide by mitosis to produce new ones. These new cells grow larger, and some of them will divide again. So growth involves both cell division and cell growth.

One way of measuring growth of an organism is to measure its mass. If you just find the mass of the organism including all the water in it, as was done to produce the graph in Fig 8.30, this is called **wet mass**. However, the amount of water in an organism can vary a lot from day to day, so it is often better to remove all the water before measuring the mass. This measurement is called **dry mass**. Obviously, the organism is killed as it is dried, so dry mass is not a good method to use for measuring the growth of animals.

Growth can be defined as *an increase in dry mass*. However, while an organism is growing, it is often becoming more complex at the same time. For example, as a seedling grows, some of its cells develop into xylem and phloem tissues, or into leaves and flowers. As a human embryo grows, it develops from a simple ball of cells into a human being with all its different tissues and organs. This increase in complexity is called **development**.

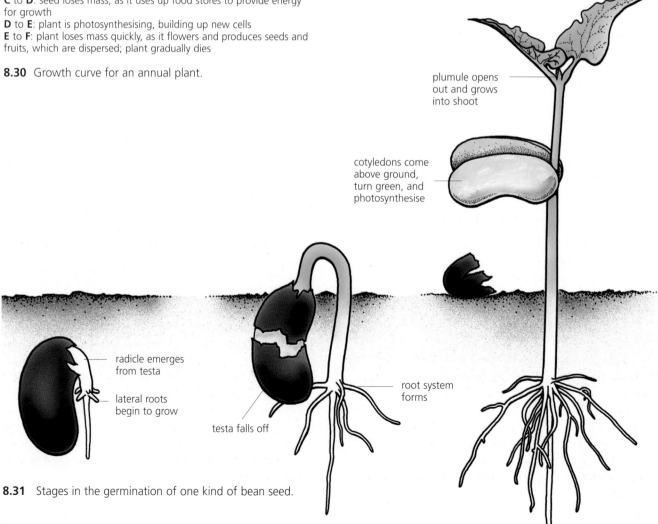

8.31 Stages in the germination of one kind of bean seed.

Practical 8.3 To find the conditions necessary for the germination of mustard seeds

1 Set up five tubes as shown in Fig 8.32.
2 Put tubes A, D and E in a warm place in the laboratory, in the light.
3 Put tube B in a refrigerator.
4 Put tube C in a warm, dark cupboard.
5 Fill in the results table to show what conditions the seeds in each tube have. The first line has been done for you.
6 Leave all the tubes for several days, then examine them to see if the seeds have germinated or not.

Results table

Tube	A	B	C	D	E
Water	✓	✓	✓	✓	✗
Warmth					
Oxygen					
Light					
Did seeds germinate?					

Questions

1 What three conditions do mustard seeds need for germination?
2 Read Sections 8.41 and 8.42, and then explain why each of these conditions is needed for successful germination.

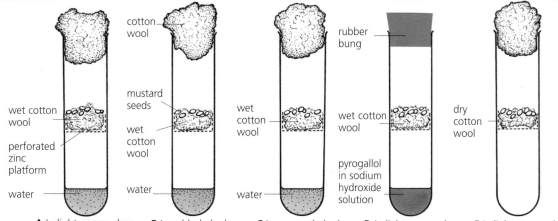

8.32 Experiment to find the conditions necessary for the germination of mustard seeds.

Questions

1 What do the cotyledons of a French bean seed contain?
2 What does dormant mean?
3 What is the advantage of dormancy?
4 What activates the enzymes in the cotyledons?
5 What do the enzymes do?

Sexual and asexual reproduction

8.44 Sexual reproduction produces variation.

This chapter has described several methods of asexual reproduction, and of sexual reproduction. There are some very important differences between them.

In asexual reproduction, some of the parent's cells divide by mitosis. This makes new cells with the same number and kind of chromosomes as the parent's cells. The new organisms are genetically identical to their parent. **Asexual reproduction does not produce variation**.

But in sexual reproduction some of the parent's cells divide by meiosis. The cells which are made are called gametes, and they have only half as many chromosomes as the parent cell. When two sets of chromosomes in two gametes combine during fertilisation, a new combination of genes is produced. The new organism will be different from either of its parents. **Sexual reproduction produces variation**.

8.45 Sexual and asexual reproduction each have advantages.

Is it useful or not to have variation amongst offspring? Sometimes, it is a good thing not to have variation. If a plant, for example, is growing well in a particular place, and if there is plenty of room for it, then it is advantageous if it produces a lot more plants just like itself. The plant is well adapted to living in these conditions and if its offspring are identical to it, then they will be well adapted, too. Also, asexual reproduction is likely to be a quicker method than sexual reproduction, because there is no need to find a mating partner.

However, if the plant is having difficulty in surviving, or if space is very limited, then sexual reproduction might be more advantageous. The seeds produced could be scattered over a wide area. The new plants which grow from them will all be slightly different from one another, and there is a good chance that some of them will be well adapted to the new conditions they find themselves in.

In general, asexual reproduction is beneficial in an unchanging environment, or when spreading out in a new area where the parent organism is well adapted to survive. Sexual reproduction is most useful in an unstable environment, where variation in the offspring might produce organisms able to survive in a variety of conditions. This helps organisms to begin to colonise new areas.

You will find more about variation, and its importance in evolution, in Chapter 12.

Chapter revision questions

1. Match each of these words with its definition.

   zygote, mitosis, meiosis, gamete, pollination, fertilisation, pericarp, fruit, seed

   (a) a sex cell, containing only half the normal number of chromosomes
   (b) an ovary after fertilisation
   (c) a diploid cell, formed by the fusion of two gametes
   (d) a type of cell division which produces daughter cells just like the parent cell
   (e) a type of cell division which produces daughter cells with only half the number of chromosomes as the parent cell
   (f) an ovary wall after fertilisation
   (g) the transfer of pollen from an anther to a stigma
   (h) an ovule after fertilisation
   (i) the fusion of two gametes

2. (a) Use the following table to plot a growth curve for a pea plant.
   (b) Explain exactly how these results would have been obtained.
   (c) Explain the reasons for the shape of the curve.

Time from planting of seed / weeks	Dry mass / g
0	1.4
0.5	0.8
1	1.6
1.5	2.5
2	5.2
3	21.5
4	31.6
5	41.3
6	49.0
8	61.1
10	66.5
12	67.4
14	66.4
15	67.2
16	66.0
17	53.2

3. Figure 8.33 shows two types of *Primula* flower. These types of flower are often found growing close together. Any one *Primula* plant, however, only has one type of flower.

   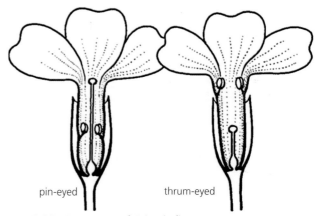

   8.33 Two types of *Primula* flower.

   (a) Describe the difference in the arrangement of the anthers and stigmas in the pin-eyed and thrum-eyed flowers.
   (b) This species is pollinated by insects, which reach into the bottom of the flower to get the nectar. Which part of the insect's body would pick up pollen in (i) a pin-eyed flower and (ii) a thrum-eyed flower?
   (c) Which part of the insect's body would touch the stigma in (i) a pin-eyed flower, and (ii) a thrum-eyed flower?
   (d) Explain how this will help to ensure that cross-pollination takes place.
   (e) Self pollination sometimes occurs in this species. Would you expect it to occur more often in pin-eyed or thrum-eyed flowers? Explain your answer.
   (f) Why is cross-pollination usually preferable to self-pollination?

Chapter revision questions (continued)

4 (a) Which type of cell division is involved in (i) the production of a new organism by asexual reproduction, (ii) the production of gametes, and (iii) the growth of a zygote?
(b) With the aid of diagrams, describe one way in which a named plant naturally reproduces asexually.
(c) What advantages are there to the plant in reproducing in this way?
(d) Many plants also reproduce sexually. What are the advantages to a plant in reproducing in this way?

5 Use the table below to plot growth curves for a human male, and a human female. Plot both curves on the same axes.

Age / years	Height / cm Male	Female
0	53.0	53.0
1	61.0	61.0
2	71.0	71.0
3	91.5	86.5
4	99.0	91.5
5	104.5	96.0
6	108.5	101.0
7	114.0	111.0
8	122.0	119.0
9	124.5	124.0
10	124.5	127.5
11	127.0	130.0
12	129.5	132.0
13	131.5	134.0
14	137.0	137.0
15	142.0	142.0
16	147.0	147.0
17	155.0	152.5
18	162.5	157.0
19	170.0	160.0
20	172.5	161.5
21	175.0	162.0
22	175.0	162.0
23	175.0	162.0
24	175.0	162.0

(a) At which age does growth appear to stop?
(b) At which age is the difference in height between male and female (i) least, and (ii) greatest?
(c) How much does the female grow between the ages of 9 and 17?
(d) What is the average rate of growth per year for the female between the ages of 9 and 17?
(e) Do you think that height is a good way of measuring human growth? Give reasons for your answer.

6 (a) State what is meant by the term sexual reproduction. [3]
(b) Fig 8.34 shows the male reproductive system.

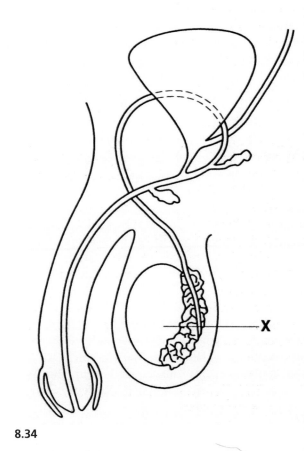

8.34

(i) Name the part labelled **X** and state two of its functions. [3]
(ii) Birth control can be brought about by surgery. Mark clearly on a copy of Fig 8.34 where such an operation would be carried out in a male. [1]

0610/2 J00

9 Coordination and response

9.1 Organisms detect changes around them.

All living organisms are sensitive to their environment. This means that they can detect changes in their environment. The changes they detect are called **stimuli**.

Cells that detect stimuli are called **receptor cells**. In animals, the receptors are often part of a **sense organ**. For example, your eye is a sense organ and the rod and cone cells in the retina of your eye are receptors. They are sensitive to light.

Table 9.1 lists some of the sense organs and receptors in the human body.

The eye

9.2 The eye is well protected.

The part of the eye which contains the receptor cells is the **retina** (Fig 9.1). This is the part which is actually sensitive to light. The rest of the eye simply helps to protect the retina, or to focus light onto it.

Each eye is set in a bony socket in the skull, called the **orbit**. Only the very front of the eye is not surrounded by bone.

The front of the eye is covered by a thin, transparent membrane called the **conjunctiva**, which helps to protect the parts behind it. The conjunctiva is always kept moist by a fluid made in the **tear glands**. This fluid contains an enzyme called **lysozyme**, which can kill bacteria.

The fluid is washed across your eye by your eyelids when you blink. The eyelids, eyebrows and eyelashes also help to stop dirt from landing on the surface of your eyes.

Even the part of the eye inside the orbit is protected. There is a very tough coat surrounding it called the **sclera**.

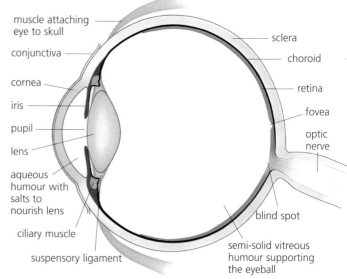

9.1 Section through a human eye.

9.3 Cells in the retina are receptive to light.

The retina is at the back of the eye. It contains two sorts of receptor cell. **Rods** are sensitive to quite dim light, but only let you see in black and white. **Cones** give colour vision, but only in bright light.

When light falls on a receptor cell in the retina, the cell sends a message along the **optic nerve** to the brain. The brain sorts out all the messages from each receptor cell, and builds up an **image**.

The closer together the receptor cells are, the clearer the image the brain will get. The part of the retina where the receptor cells are packed most closely together is called the **fovea**. This is the part of the retina where light is focused when you look straight at an object. All the receptor cells in the fovea are cones. The rods are scattered further out on the retina.

Table 9.1 Sense organs and receptors in the human body.

Sense organ	Receptor cells	Stimulus to which the receptors respond
Eye	Rods and cones	Light
Ear	Hair cells in the cochlea	Sound
Skin	Nerve endings in the dermis	Touch and temperature
Nose	Cells in the lining of the nasal passages	Chemicals
Tongue	Cells in the taste buds	Chemicals

There are no receptor cells where the optic nerve leaves the retina. This part is called the **blind spot**. If light falls on this place, no messages will be sent to the brain.

Behind the retina is a black layer called the **choroid**. The choroid absorbs all the light after it has been through the retina, so it does not get scattered around the inside of the eye.

9.4 The eye focuses light.

For the brain to see a clear image, there must be a clear image focused on the retina. Light rays must be bent, or **refracted**, so that they focus exactly onto the retina.

The **cornea** is responsible for most of the bending of the light. The **lens** makes fine adjustments. Fig 9.2 shows how these two parts of the eye focus light onto the retina. The image on the retina is upside down. The brain interprets this so that you see it the right way up.

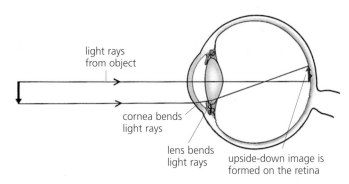

9.2 How an image is focused onto the retina.

9.5 The lens adjusts the focusing.

Not all light rays need bending the same amount to focus them onto the retina. Light rays coming from a nearby object are spreading apart from one another, or diverging. They will need to be bent inwards quite strongly.

Light rays coming from an object in the distance will be almost parallel to one another. They will not need bending so much.

The shape of the lens can be adjusted to bend light rays more. The fatter it is, the more it will bend them. The thinner it is, the less it will bend them. The adjustment in the shape of the lens, to focus light coming from different distances, is called **accommodation**.

Figs 9.3, 9.4 and 9.5 show how the shape of the lens is changed. It is held in position by a ring of **suspensory ligaments**. The tension on the suspensory ligaments, and thus the shape of the lens, is altered by means of the **ciliary muscle**. When this ring of muscle contracts, the suspensory ligaments are loosened. When the ring of muscle relaxes, they are pulled tight. When the suspensory ligaments are tight, the lens is pulled thin. When they are loosened, the lens gets fatter.

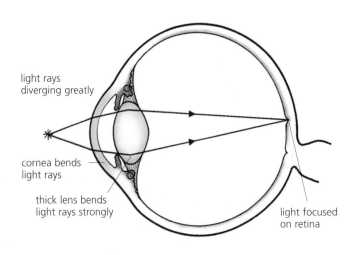

9.3 Focusing on a nearby object.

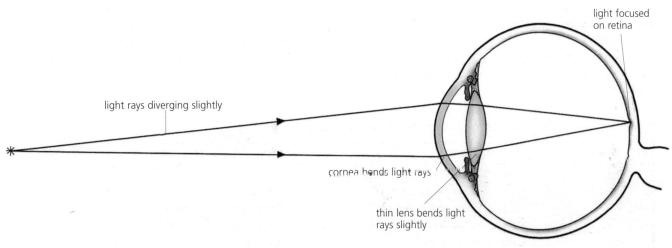

9.4 Focusing on a distant object.

Practical 9.1 Looking at human eyes.

It is best to perform this experiment with a partner, although it is possible to use a mirror and look at your own eyes.

1. First identify all the following structures: eyebrows; eyelashes; eyelids; conjunctiva; pupil; iris; cornea; sclera; small blood vessels; openings to tear ducts. Fig 9.1 will help you to do it.
2. Make a diagram of a front view of the eye and label each of these structures on it.
3. Use Sections 9.2 to 9.6 to find out the functions of each structure you have labelled. Write down these functions, as briefly as you can, next to each label or beneath your diagram.
4. Ask your partner to close his or her eyes, and cover them with something dark to cut out as much light as possible. (Alternatively, you may be able to darken the whole room.) After about 3 or 4 minutes, quickly remove the cover (or switch on the lights) and look at your partner's eyes as they adapt to the light. What happens? What is the purpose of this change?
5. Read Section 9.6, and then explain how this change is brought about.

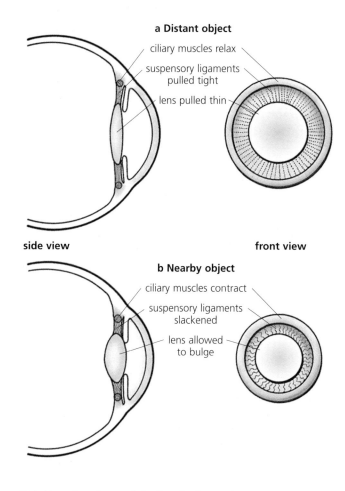

9.5 How the shape of the lens is changed.

Practical 9.2 Dissecting a sheep's eye.

1. Carefully examine the eye. Using forceps and a scalpel (ask to be shown how to use them correctly), remove as much of the white **fat** as you can. Be careful, though, not to damage the brownish coloured **muscles** attached to the outside of the eye, or the white **optic nerve** which comes out at the back of it.
2. Draw the eye, and label: conjunctiva and cornea; iris; sclera; fat; eye muscles; optic nerve; pupil.
3. Using sharp scissors, make a small incision into the eye as shown. What comes out? What happens to the shape of the eye? So what is one of the functions of this substance?
4. Continue cutting around the eye until you have cut it completely in half.
5. First, look at the back half. The **retina** may have detached itself from here, and may have floated away in the fluid. The next layer in is the black **choroid**. Is it still there? What is the function of the choroid?
6. Behind the choroid is the **sclera**. What is it like? What is its function?
7. Now investigate the front half of the eye. The **lens** will probably be floating loose. What normally holds the lens in position? What does the lens look like? If the lens is not too cloudy, put it over some writing and look through it. What does it do?
8. Try to find other structures at the front of the eye, for example the **iris**. Identify and describe any structures you can find.

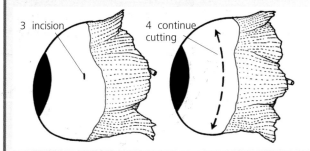

9.6 The iris adjusts how much light enters the eye.

In front of the lens is a circular piece of tissue called the **iris**. The iris contains pigments (coloured substances), which absorb light and stop it getting through to the retina.

In the middle of the iris is a space, called the **pupil**. The size of the pupil can be adjusted. The wider the pupil is, the more light can get through to the retina. In strong light, the iris closes in, and makes the pupil small. This stops too much light getting in and damaging the retina.

The iris contains muscles which allow it to adjust the size of the pupil. **Circular muscles** lie in circles around the pupil. When they contract, they make the pupil constrict, or get smaller. **Radial muscles** run outwards from the edge of the pupil. When they contract, they make the pupil dilate, or get larger. This response of the iris to a change in light intensity is an example of a **reflex action** (Section 9.12).

Coordination

9.7 Effectors respond when the body is stimulated.

Being able to detect stimuli is not much use to an organism unless it can respond to them in some useful way. The part of the body which responds to a stimulus is called an **effector**.

Muscles are effectors. For example, if you touch something hot, the muscles in your arm contract, so that your hand is quickly pulled away. Glands can also be effectors. If you smell food cooking, your salivary glands may react by secreting saliva.

To make sure that the right effectors respond at the right time, there needs to be some kind of communication system between receptors and effectors. The pain receptors on your fingertips need to send a message to your arm muscles to tell them to contract. The chemical receptors in your nose must communicate with your salivary glands, to make them secrete saliva. The way in which receptors pick up stimuli, and then pass messages on to effectors, is called **coordination**.

9.8 Hormones and nerves allow communication.

Animals need fast and efficient communication systems between their receptors and effectors. This is partly because most animals move in search of food. Many animals need to be able to respond very quickly to catch their food, or to avoid predators.

Most animals have two methods of sending messages from receptors to effectors. The fastest is by means of **nerves** (Fig 9.7, overleaf). The receptors and nerves make up the animal's **nervous system**. A slower method, but still a very important one, is by means of chemicals called **hormones**. Hormones are part of the **endocrine system**.

Nervous systems

9.9 Neurones carry nerve impulses.

Nervous systems are made of special cells called **neurones**. Fig 9.6 (overleaf) illustrates a neurone from a mammal's body.

Neurones contain the same basic parts as any animal cell. Each has a nucleus, cytoplasm, and a cell membrane. But their structure is specially adapted to be able to carry messages very quickly.

To enable them to do this, they have long, thin fibres of cytoplasm stretching out from the cell body. They are called **nerve fibres**. The longest fibre in Fig 9.6 is called an **axon**. Axons can be more than a metre long. The shorter fibres are called **dendrons** or **dendrites**.

The dendrites pick up messages from other neurones lying nearby. They pass the message to the cell body, and then along the axon. The axon might then pass it on to another neurone.

> ### Questions
> 1. What is a stimulus?
> 2. Name two parts of the body which contain receptors of chemical stimuli.
> 3. Which part of the eye contains cells which are sensitive to light?
> 4. Your brain can build up a very clear image when light is focused onto the fovea. Explain why it can do this.
> 5. If you look straight at an object when it is nearly dark, you may find it difficult to see it. It is easier to see if you look just to one side of it. Explain why this is.
> 6. What is the choroid, and what is its function?
> 7. List, in order, the parts of the eye through which light passes to reach the retina.
> 8. Name two parts of the eye which refract light rays.
> 9. What is meant by accommodation?
> 10. (a) What do the ciliary muscles do when you are focusing on a nearby object?
>   (b) What effect does this have on (i) the suspensory ligaments, and (ii) the lens?

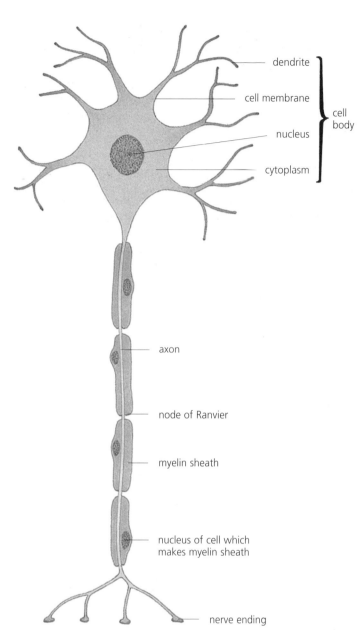

9.6 A motor neurone.

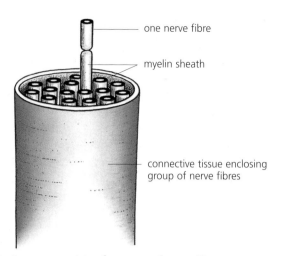

9.7 A nerve consists of a group of nerve fibres.

9.11 Most animals have a central nervous system.

The **brain** and **spinal cord** make up the **central nervous system**, or **CNS** (Fig 9.8). Like the rest of the nervous system, the CNS is made up of neurones. Its job is to coordinate the messages travelling through the nervous system.

When a receptor detects a stimulus, it sends an electrical impulse to the brain or spinal cord. The brain or spinal cord receives the impulse, and 'decides' which effectors need to react to the stimulus. It then sends an impulse on, along the appropriate nerve fibres, to the appropriate effector.

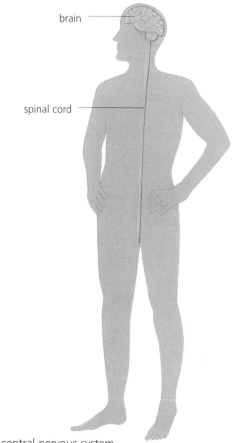

9.8 The human central nervous system.

9.10 Myelinated neurones carry impulses quickly.

Some of the nerve fibres of active animals like mammals are wrapped in a layer of fat and protein called **myelin**. Every now and then, there are narrow gaps in the myelin sheath.

The messages that neurones transmit are in the form of electrical impulses. Myelin insulates the nerve fibres, so that they can carry impulses much faster. A myelinated nerve fibre in a cat's body can carry impulses at up to 100 metres per second. A fibre without myelin can only carry impulses at about 5 metres per second.

9.12 Reflex arcs allow rapid response.

Fig 9.9 shows how these impulses are sent. If your hand touches a hot plate, an impulse is picked up by a sensory receptor in your finger. It travels to the spinal cord along the axon from the receptor cell. This cell is called a **sensory neurone**, because it is carrying an impulse from a sensory receptor.

In the spinal cord, the neurone passes an impulse on to several other neurones. Only one is shown in Fig 9.9. These neurones are called **relay neurones**, because they relay the impulse on to other neurones. The relay neurones pass the impulse on to the brain. They also pass it on to an effector.

In this case, the effectors are the muscles in your arm. The message travels to the muscle along the axon of a **motor neurone**. The muscle then contracts, so that your hand is pulled away.

This sort of reaction is called a **reflex action**. You do not need to think about it. Your brain is made aware of it, but you only consciously realise what is happening after the message has been sent on to your muscles.

Reflex actions are very useful, because the message gets from the receptor to the effector as quickly as possible. You do not waste time in thinking about what to do. The arrangement of sensory neurone, relay neurones and motor neurone is called a **reflex arc**.

Another example of a reflex action is the response of the iris to a change in light intensity. Imagine that you walk from a bright, sunny street into a dark room. The receptor cells in the retinas of your eyes sense that less light is falling onto them, and they send a nerve impulse along the optic nerve to the brain. Here, one or more relay neurones pick up the signal, and send nerve impulses along a motor neurone to the muscles in the iris. The radial muscles in the iris respond to the impulse by contracting, while the circular muscles relax. This makes the width of the iris smaller, so that it covers less of the pupil and more light is allowed into the eye.

9.13 Synapses connect neurones.

If you look carefully at Fig 9.9, you will see that the three neurones involved in the reflex arc do not quite connect with one another. There is a small gap between each pair. These gaps are called **synapses**. Fig 9.10 shows a synapse in more detail. Inside the sensory neurone's axon are hundreds of tiny vacuoles, or **vesicles**. These each contain a chemical, called **transmitter substance**.

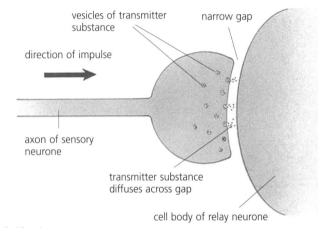

9.10 A synapse.

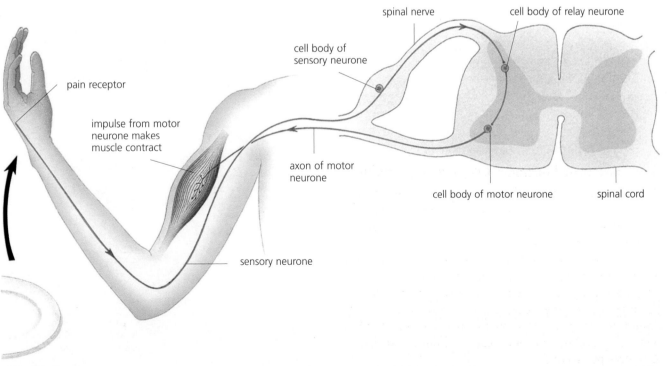

9.9 A reflex arc.

When an impulse comes along the axon, it makes these vesicles empty their contents into the space between the two neurones. The transmitter substance quickly diffuses across the tiny gap, attaches to the membrane of the relay neurone and triggers an impulse in the relay neurone. The relay neurone then sends the message onwards.

Synapses act like one-way valves. There is only transmitter substance on one side of the gap, so messages can only go across from that side. Synapses ensure that nervous impulses only travel in one direction.

Practical 9.3 Measuring reaction time

The time taken for a nerve impulse to travel from a receptor, through your CNS and back to an effector is very short. It can be measured, but only with special equipment. However, you can get a reasonable idea of the time it takes if you use a large number of people, and work out an average time.

1 Get as many people as possible to stand in a circle, holding hands.
2 One person lets go of his or her neighbour with the left hand, and holds a stopwatch in it. When everyone is ready, this person simultaneously starts the stopwatch, and squeezes the neighbour's hand with the right hand.
3 As soon as each person's left hand is squeezed, he or she should squeeze his or her neighbour with the right hand. The message of squeezes goes all round the circle.
4 While the message is going round, the person with the stopwatch puts it into the right hand, and holds his or her neighbour's hand with the left hand. When the squeeze arrives, he or she should stop the watch.
5 Keep repeating this, until the message is going round as fast as possible. Record the time taken, and the number of people in the circle.
6 Now try again, but this time make the message of squeezes go the other way around the circle.

Questions

1 Using the fastest time you obtained, work out the average time it took for one person to respond to the stimulus they received.
2 Did people respond faster as the experiment went on? Why might this happen?
3 Did the message go as quickly when you changed direction? Explain your answer.

9.14 A complex CNS allows complex behaviour.

Why is the central nervous system needed? Would it not be much quicker if pain receptors in your hand could just send a message straight to your arm muscles to tell them to move your hand away from the hot plate, rather than all the way to the spinal cord and back? Yes, it would, but that system would not be good enough for animals which need to be able to vary their behaviour under different circumstances.

With a central nervous system it is possible to give a modified, more 'intelligent' response. Say, for example, that you started to pick up the hot plate before you knew it was hot. If you just pulled your hand away, you would drop the plate and break it.

When the message from your fingers saying 'hot plate' arrives at your CNS, there is already another message there saying 'but don't drop it'.

The CNS will 'consider' the two messages together. It will probably send a message to your muscles to tell them to put the plate down gently, not to drop it.

The job of the CNS is to collect up all the information from all the receptors in your body. This information will be added together before messages are sent to effectors. In this way, the best action can be taken in a particular set of circumstances.

9.15 Brain functions are localised.

The brain and spinal cord both help to receive impulses from receptors, and pass them on to effectors. But the brain does much more than this.

Figs 9.11 and 9.12 show the structure of the human brain. It is surrounded by three membranes or **meninges**, which help to protect it.

The **cerebrum** is the largest part of the brain. It is made of two **cerebral hemispheres**. Mammals have much larger cerebral hemispheres than any other kind of animal. Humans have the largest ones of all, compared with the size of the rest of the brain.

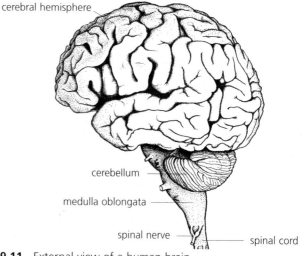

9.11 External view of a human brain

Conscious thought and memory take place in the cerebrum. Different parts of the cerebrum have different functions. For example, some areas deal with sight, others with speech. An area near the front determines some aspects of your personality.

The **hypothalamus** lies underneath the front part of the cerebrum. This is the part of the brain which controls osmoregulation and temperature regulation (Chapter 10). The **cerebellum** is in control of coordination of body movements, and posture. The **medulla oblongata**, sometimes simply known as the medulla, controls heart beat and breathing.

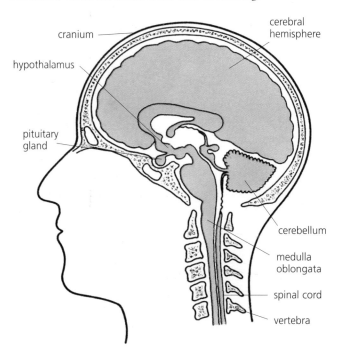

9.12 Section through a human head, to show the brain.

9.16 Human actions can be voluntary or involuntary.

We have seen how some of our responses to stimuli happen very quickly, without having to think about it. Actions that we do without thinking consciously about them are called involuntary actions.

Reflex actions are always involuntary. Some reflex actions seem to be 'hard-wired' into our brains, so that they take place even in very young children and do not have to be learned. An example of this kind of reflex is the iris reflex, described in Section 9.12. Other reflex actions are learned. An example of this kind of reflex is making the many tiny adjustments to the contraction of muscles in your legs, back and arms that allow you to ride a bicycle without falling off.

But many human actions are voluntary ones. A voluntary action is one over which we have conscious control. An example of a voluntary action is responding to the sight of friends by calling out to them.

You make a decision about whether or not to do this. Another example is picking up a piece of food and putting it into your mouth.

9.17 Invertebrate behaviour is often very simple.

As far as we know, invertebrates do not think about their actions in anything like the ways that we do. Studies of invertebrate behaviour show that, most of the time, a particular stimulus will always produce the same response. These responses have evolved (Chapter 12) because they help the animal to survive in its environment.

For example, some species of flies lay their eggs on the bodies of dead animals. When the eggs hatch, small larvae called maggots hatch out. They feed on the decaying flesh. They spend most of their time either inside or underneath the decaying carcase. This hides them from predators that might eat them, and it also helps their soft bodies not to dry out.

Fly maggots do not think to themselves 'I must bury myself underneath this carcase'. They react instinctively and automatically to light by moving away from it. If you place a maggot on a piece of paper and shine a light towards it, the maggot will move its head from side to side. It moves its body around until the light is shining equally on both sides of its head, and then it moves directly away from the direction in which the light is coming. This is called **negative phototaxis**. It is negative because the maggots move away from the light. 'Photo' means 'light', and 'taxis' means 'movement'.

Questions

1. Give two examples of effectors.
2. What are the two main communication systems in an animal's body?
3. List three ways in which neurones are similar to other cells.
4. List three ways in which neurones are specialised to carry out their function of transmitting messages very quickly.
5. What is a nerve?
6. What is the function of the central nervous system?
7. Where are the cell bodies of each of these types of neurone found: (a) sensory neurone, (b) relay neurone, and (c) motor neurone?
8. Apart from where they are found, describe the differences in appearance between a motor neurone and a sensory neurone.
9. What is the value of reflex actions?

The endocrine system

9.18 Endocrine glands make hormones.

Nerves can carry electrical messages very quickly from one part of an animal's body to another. But animals also use chemical messages.

The chemicals are called **hormones**. Hormones are made in special glands called **endocrine glands**. A hormone can be defined as a chemical substance, produced by a gland and carried by the blood, which alters the activity of target cells in one or more specific organs.

Endocrine glands have a good blood supply. They have blood capillaries running right through them. When the endocrine gland makes a hormone, it releases it directly into the blood.

Other sorts of gland do not do this. The salivary glands, for example, do not secrete saliva into the blood. Saliva is secreted into the salivary duct, which carries it into the mouth. Endocrine glands do not have ducts, so they are sometimes called ductless glands.

Once the hormone is in the blood, it is carried to all parts of the body, dissolved in the plasma. Each kind of hormone only affects certain parts of the body. Hormones are eventually broken down in the liver or lost in the urine.

Fig 9.13 shows the position of some endocrine glands in the human body, while Table 9.2 summarises the hormones which they secrete, and their effects. The functions of adrenaline, and the male and female sex hormones are described in the next few pages. The functions of insulin and glucagon are described in Chapter 10.

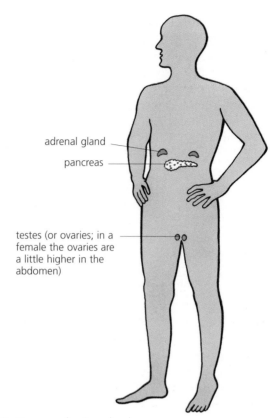

9.13 Some endocrine glands.

Table 9.2 Mammalian endocrine glands and hormones.

Hormone	Gland which secretes it	When secreted	Function	Other points
Adrenaline	Adrenal gland	In small amounts all the time; in large amounts when frightened	Prepares the body for fight or flight	
Insulin	Islets of Langerhans in pancreas	When blood glucose level rises above normal	Causes liver and muscles to take up glucose and convert it to glycogen, so restoring blood glucose level to normal	Lack of insulin causes diabetes
Glucagon	Islets of Langerhans in pancreas	When blood glucose drops below normal	Causes liver to break down glycogen and release glucose into the blood	
Testosterone	Testes	In small quantities throughout life; in larger quantities from puberty onwards	Controls development of male sex organs and secondary sexual characteristics	
Oestrogen	Ovaries	In small quantities throughout life; in larger quantities from puberty onwards, particularly when follicle is developing in ovary	Controls the development of sex organs and secondary sexual characteristics; causes lining of uterus to get thick and spongy	
Progesterone	Corpus luteum	After ovulation	Maintains lining of uterus	If placenta does not secrete enough progesterone, a miscarriage may occur
	Placenta	Throughout pregnancy		

9.19 Adrenaline prepares the body for action.

There are two adrenal glands, one above each kidney. They make a hormone called **adrenaline**. When you are frightened, excited or keyed up, your brain sends impulses along a nerve to your adrenal glands. This makes them secrete adrenaline into the blood.

Adrenaline has several effects which are designed to help you to cope with danger. For example, it makes your heart beat faster, supplying oxygen to your brain and muscles more quickly. This gives them more energy for fighting or running away.

The blood vessels in your skin and digestive system become narrow so that they carry very little blood. This makes you go pale, and gives you 'butterflies in your stomach'. As much blood as possible is needed for your brain and muscles in the emergency.

Adrenaline also stimulates the liver to break down glycogen and release glucose into the blood. This makes sure that muscle cells get plenty of glucose to use in respiration, to provide energy so they can contract.

All of this is very useful if you really have to fight an enemy. It is also useful if you are an athlete at the start of a race. But it does not help at all if you are on your way to the dentist, or watching a horror film.

Like most hormones, adrenaline breaks down very quickly after it is released, so its effects do not last long. If you need to go on feeling frightened, then your brain will keep telling the adrenal glands to secrete more adrenaline.

Table 9.2 A comparison of the nervous and endocrine systems in a mammal.

Nervous system	Endocrine system
Made of neurones	Made of secretory cells
Messages transmitted in the form of electrical impulses	Messages transmitted in the form of chemicals called hormones
Messages transmitted along nerve fibres	Messages transmitted through the blood system
Messages travel very quickly	Messages travel more slowly
Effect of message usually only lasts a very short while	Effect of messages usually lasts longer

9.20 Male sex hormones.

Male sex hormones are called **androgens**. The most important androgen is **testosterone**. Testosterone is made in the testes.

Testosterone and other androgens regulate the development of the male sex organs. They also control the development of the male secondary sexual characteristics (Section 8.26).

9.21 Female sex hormones control the menstrual cycle.

Female sex hormones are called **oestrogens**. They regulate the development of the female sex organs, and the female secondary sexual characteristics.

Whereas male mammals make sperm all the time, females only produce eggs at certain times. In humans, ovulation (Section 8.14) happens once a month. Ovulation is part of the **menstrual cycle**. The menstrual cycle is controlled by hormones (Fig 9.14).

First, a follicle develops inside an ovary. The developing follicle secretes a hormone called **oestrogen**. The oestrogen makes the lining of the uterus grow thick and spongy.

When the follicle is fully developed, ovulation takes place. The follicle stops secreting oestrogen. It becomes a **corpus luteum**. The corpus luteum starts to secrete another hormone, called **progesterone**.

Progesterone keeps the uterus lining thick, spongy, and well supplied with blood, in case the egg is fertilised. If it is not fertilised, then the corpus luteum gradually disappears. Progesterone is not secreted any more, and so the lining of the uterus breaks down. Menstruation happens. A new follicle starts to develop in the ovary, and the cycle begins again.

But if the egg is fertilised, the corpus luteum does

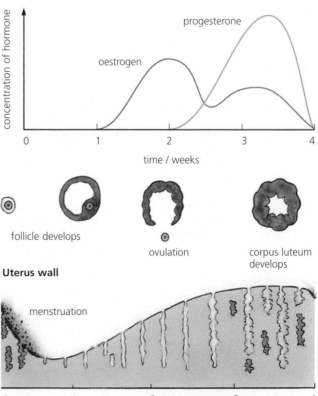

9.14 Hormones and the menstrual cycle.

not degenerate so quickly. It carries on secreting progesterone until the embryo sinks into the uterus wall, and a placenta develops. Then the placenta secretes progesterone, and carries on secreting it all through the pregnancy. The progesterone maintains the uterus lining, so that menstruation does not happen during the pregnancy.

> **Questions**
> 1 Which female hormone is secreted by a follicle as it develops inside an ovary?
> 2 What effect does this hormone have?
> 3 Which hormone is secreted by a corpus luteum?
> 4 What effect does this hormone have?
> 5 Name one other structure which secretes this hormone.

9.22 Animal hormones are sometimes used in food production.

Farmers sometimes use hormones to make their animals grow faster, or to produce more of a particular product. One hormone used in this way is called **bovine somatotropin**, or **BST**.

BST is a hormone which is naturally produced by cattle. However, if cows are given *extra* BST, they make more milk. Some people think it would be a good idea to give cows BST, to get higher milk yields. You would need fewer cows to get the same amount of milk.

There are several arguments against it:
- **Some people are worried about drinking milk from cows treated with BST.**
  They think the BST might damage their health. In fact, this is very unlikely, because the hormone does not get into the milk in any significant quantity.
- **It is difficult to see why we need BST.**
  The European Union already produces more milk than it needs, so milk quotas have to be imposed, to stop farmers from producing too much milk.
- **There are concerns that the BST might harm the cows.**
  Cows treated with BST make very large amounts of milk, far beyond the 'natural' levels which they produce. This makes them more likely to get infections of their udders, and may make them feel less comfortable.

Coordination in plants

9.23 Tropisms are directional growth responses.

Most plants cannot respond to stimuli as quickly as animals can. They respond more slowly, usually by growing. They grow either towards or away from the stimulus. This sort of response is called a **tropism**.

Two important stimuli for plants are light and gravity. For example, the shoot of a plant grows towards light. This is called **phototropism** ('photo' means light). Because the shoot grows *towards* the light, the response is called **positive phototropism**.

Plants can respond to gravity by growing either towards or away from the centre of the Earth. This is called **geotropism**. Shoots tend to grow away from the pull of gravity. This is called **negative geotropism** because the shoot is growing *away from* the stimulus.

Roots are **positively geotropic** – they grow towards the pull of gravity (Fig 9.15). Some roots also respond to light by growing away from it, which means that they are **negatively phototropic**.

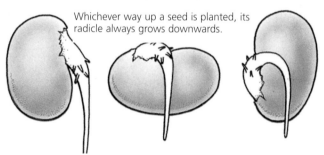

9.15 Positive geotropism in roots.

9.24 Tropisms aid plant survival.

It is very important to the plant that its roots and shoots behave like this. Shoots must grow upwards, away from gravity and towards the light, so that the leaves are held out into sunlight. The more light they have, the better they can photosynthesise. Flowers, too, need to be held up in the air, where insects or the wind can pollinate them.

Roots, though, need to grow downwards into the soil in order to anchor the plant in the soil, and to absorb water and minerals from between the soil particles.

9.25 Shoots responds to light.

For an organism to respond to a stimulus, there must be a **receptor** to pick up the stimulus, an **effector** to respond to it, and some kind of **communication system** in between. In mammals, the receptor is often part of a sense organ, and the effector is a muscle or gland. Messages are sent between them along nerves, or sometimes by means of hormones.

Plants, however, do not have complex sense organs, muscles or nervous systems. So how do they manage to respond to stimuli like light and gravity?

The part of a shoot sensitive to light is the tip. This is where the receptor is. The part of the shoot which responds to the stimulus is the part just below the tip. This is the effector.

These two parts of the shoot must be communicating with one another somehow. They do it by means of **hormones**.

Plant hormones, like animal hormones, are chemicals that affect the activities of particular cells and organs. However, they are not made in glands, and they are not transported in blood! Plant hormones may move from one part of the plant to another in the phloem, or by diffusion, or by active transport from cell to cell.

9.26 Changes in auxin concentration cause phototropisms.

One kind of plant hormone is called **auxin**. Auxin is being made all the time by the cells in the tip of a shoot. The auxin diffuses downwards from the tip, into the rest of the shoot. Auxin makes the cells just behind the tip get longer. The more auxin there is, the faster they will grow. Without auxin, they will not grow.

When light shines onto a shoot from one side, the auxin at the tip concentrates on the shady side (Fig 9.16). This makes the cells on the shady side grow faster than the ones on the bright side, so the shoot bends towards the light.

9.27 Plant hormones are used in food production.

Gardeners and horticulturists often use plant hormones to improve the look of their gardens, to increase yields from plants, or to speed up the rate at which they can produce new plants.

Many people use weedkillers in their gardens. Most weedkillers contain plant hormones. These hormones are often a type of **auxin**. The weedkillers used to kill weeds in lawns are **selective weedkillers**. When they are sprayed onto the lawn, the weeds are affected by the auxin, but the grass is not. The weeds respond by growing very fast. Then the weeds die, leaving more space, nutrients and water for the grass to grow.

Farmers use similar weedkillers to kill weeds growing in cereal crops such as wheat, millet, maize or sorghum.

Fruit growers often use plant hormones to help the fruits to grow larger, or to ripen well. For example, many fruits produce the gas **ethene** when they are ripening. This encourages other fruits near them to ripen, as well. (There is no reason why a gas cannot be a hormone.) If tomatoes are picked while they are still green, they can be stored or transported without worrying about them going bad. When the suppliers want them to ripen, they can expose them to ethene.

Plant hormones are also used when propagating plants by cuttings or tissue culture. You can read about this in Section 8.7.

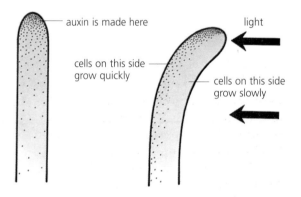

Auxin made in the tip diffuses unevenly down the shoot, concentrating on the shady side.

The uneven concentration of auxin causes the shady side to grow faster than the light side, so the shoot bends towards the light.

9.16 Auxin and phototropism.

Questions
1. What is a tropism?
2. Which parts of a plant show positive phototropism?
3. Which parts of a plant show positive geotropism?
4. Which part of a shoot is sensitive to light?
5. Which part of a shoot responds to light?

Practical 9.4 To find out how shoots respond to light

1. Label three petri dishes A, B and C. Line each with moist cotton wool or filter paper, and sprinkle on some mustard seeds.
2. Leave all three dishes in a warm place for a day or two, until the seeds begin to germinate. Check that they do not dry out.
3. Now put dish A into a light-proof box with a slit in one side, so that the seedlings get light from one side only.
4. Put dish B onto a **clinostat** (Fig 9.17) in a light place. The clinostat will slowly turn the seedlings around, so that they get light from all sides equally.
5. Put dish C into a completely light-proof box.
6. Leave all the dishes for a week, checking that they do not dry out.
7. Then, make labelled drawings of one seedling from each dish.

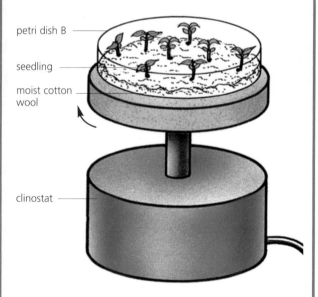

9.17 Apparatus to find out how shoots respond to light.

Questions

1. How had the seedlings in A responded to light from one side? What is the name for this response?
2. Why was dish B put onto a clinostat, and not simply left in a light place?
3. Explain what happened to the seedlings in dish C.
4. What was the control in this experiment?

Practical 9.5 To find out how roots respond to gravity

1. Germinate several broad bean seeds. Leave them until their radicles are about 2 cm long.
2. Line the containers of two clinostats with blotting paper. Dampen the paper.
3. Cover the cork discs on the clinostats with wet cotton wool.
4. Choose about eight bean seeds with straight radicles. Pin four onto each disc, with their radicles pointing straight outwards (Fig 9.18). Put the containers lined with blotting paper over them.
5. Turn both clinostats on their sides. Switch on clinostat B.
6. Leave both clinostats for a few days. Check the beans occasionally to see that they have enough water.
7. Then, make labelled drawings of the bean seedlings from each clinostat.

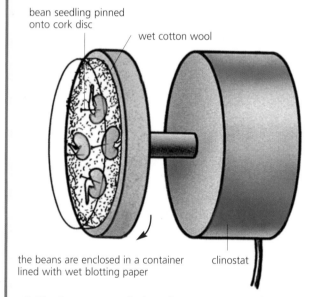

9.18 Apparatus to find out how roots respond to gravity.

Questions

1. How had the seedlings in clinostat A responded to the pull of gravity? What is the name for this response?
2. What was the purpose of clinostat B?
3. Design an experiment to find out how shoots respond to gravity. Give full instructions for the method, including labelled diagrams. What results would you expect?

Chapter revision questions

1. Describe two reflex actions, other than the one described in Section 9.12.

2. (a) How many synapses are there in the reflex arc shown in Fig 9.9?
   (b) Explain how a nerve impulse crosses a synapse.
   (c) Describe the advantages of having synapses, rather than direct connections between neurones.

3. Explain the difference between each of the following pairs of terms, giving examples whenever they make your answer clearer.
   (a) cornea, conjunctiva
   (b) choroid, sclera
   (c) neurone, nerve
   (d) receptor, effector
   (e) sensory neurone, motor neurone
   (f) cerebrum, cerebellum
   (g) oestrogen, progesterone
   (h) phototaxis, phototropism
   (i) negative geotropism, positive geotropism.

4. If you walk from a brightly lit street into a dark room, your pupils will rapidly dilate.
   (a) What type of action is this?
   (b) Using the following words at least once, but not necessarily in this order, explain how this reaction is brought about:
   synapse, receptor, motor neurone, sensory neurone, relay neurone, radial muscles.
   (c) As well as the muscles in the iris, the eye also contains muscles in the ciliary body. What is their function?

5. Fig 9.19 shows a section through the eye of a person who is focusing on a distant object.

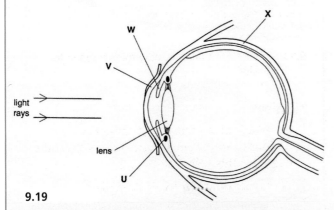

9.19

(a) Name the structures labelled U to X. [2]
When the person focuses on a book the lens of the eye changes shape.

(b) (i) Choose the lens, A to D in Fig 9.20, which most closely resembles the appearance of the lens in the eye when focussed on the book. [1]

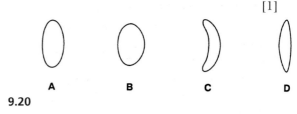

9.20

(ii) Describe what has happened in the eye to bring about this change in the shape of the lens. [4]

5090/2 N00

6. Fig 9.21 shows an experiment in which the coleoptiles (shoots) of similar seedlings have been treated in different ways.

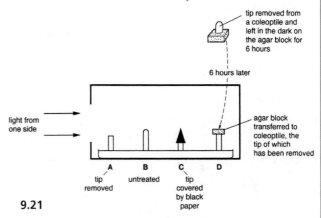

9.21

In Fig 9.22, the result in shoot D is shown 24 hours later.

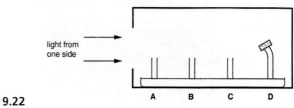

9.22

(a) (i) Name the response shown by shoot D. [1]
    (ii) Explain what has caused this response. [3]

(b) Copy and complete Fig 9.22 to show the likely results for shoots A, B and C.

(c) (i) What name is given to the simple behavioural response shown by invertebrates to external stimuli? [1]
    (ii) Many invertebrates move towards damp and dark conditions when given a choice. Suggest how this response may help them survive. [4]

0610/3 J95

10 Homeostasis and excretion

10.1 Homeostasis keeps the internal environment constant.

The environment (surroundings) of a living organism is always changing. Think about your own environment. The temperature of the air around you changes. For example, it might be −10 °C outside on a cold day in winter, and 23 °C indoors. The amount of water in the air around you changes. On a rainy day, the air could hold a lot of water, while on a hot, dry sunny day there could be very little water in the air.

The cells inside your body, however, do not have a changing environment. Your body keeps the environment *inside* your body almost the same, all the time. In the tissue fluid surrounding your cells, the temperature and amount of water are kept almost constant. So is the concentration of glucose. Keeping this internal environment constant is called **homeostasis.**

Homeostasis is very important. It helps your cells to work as efficiently as possible. Keeping a constant temperature of around 37 °C helps enzymes to work. Keeping a constant amount of water means that your cells are not damaged by absorbing or losing too much water by osmosis. Keeping a constant amount of glucose means that there is always enough fuel for respiration.

In this chapter, you will see how homeostasis is carried out in humans. The nervous system and various endocrine glands are involved, and so are the kidneys.

The control of body temperature

10.2 Mammals and birds are homeothermic.

Mammals and birds are able to keep their body temperature constant, no matter what the temperature of their environment is. They are **homeothermic.**

Being homeothermic has many advantages. You may have noticed that, in cold weather, invertebrate animals such as houseflies become very slow-moving. This is because they do not control their body temperature – they are **poikilothermic** (Fig 10.1).

All invertebrates, and also all fish, amphibians and reptiles, are poikilothermic. Their body temperature is just the same as the temperature of the air or water around them. When this temperature is cold, their body temperature is cold. Cold temperatures slow down chemical reactions (this is explained in Section 3.5), which slows down the activity of the organism.

But homeothermic animals keep their body temperature warm even when the weather is very cold. Their enzymes can carry on working, and they remain active. So mammals and birds can keep active in winter, or during cold nights, when other animals are inactive.

There is a price to pay. The energy which homeothermic animals use to generate the heat to

At 0 °C, a poikilothermic animal's metabolic rate slows down, because its body temperature is also 0 °C. The animal is inactive.

At 20 °C, a poikilothermic animal's body temperature is 20 °C. Its metabolic rate speeds up, and it becomes active.

At 0 °C, a homeothermic animal remains active. Its cells produce heat by breaking down food through respiration. Its body temperature stays high enough to keep its metabolism going.

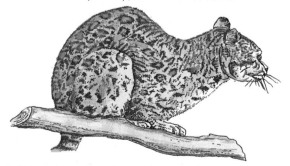

At 20 °C, a homeothermic animal is no more active than at 0 °C, because its body temperature does not change. It may even be less active, to avoid overheating.

10.1 Poikilothermic and homeothermic animals.

keep their bodies warm comes from food. This means that homeothermic animals have to eat much more food than poikilothermic ones.

10.3 Humans control their body temperature.

If you look back at Fig 9.12, you can see that there is an area at the base of the brain – almost in the middle of your head – called the **hypothalamus.** The hypothalamus keeps a constant check on the temperature of the blood flowing through it. It also receives nerve impulses from receptors in your skin which are sensitive to the temperature of your environment.

If the temperature of your blood is too low, the hypothalamus sends impulses along nerves to various parts of your body to increase heat production, and to reduce heat loss from your skin. If the temperature is too high, then impulses are sent to reduce heat production, and increase heat loss. The system is very efficient, and – as long as you are well – your body temperature rarely varies by more than 0.5 °C or so.

10.4 Human skin structure is related to function.

Fig 10.2 shows a section through human skin. Skin has many functions, one of which is to increase or decrease the rate at which heat is lost from your body to the air, which helps with temperature regulation.

Human skin is made up of two layers. The top layer is called the **epidermis**, and the lower layer is the **dermis**.

10.5 The epidermis protects the deeper layers.

All the cells in the epidermis were made in the layer of cells at the base of it – the **Malpighian layer**. These cells are always dividing by mitosis. The new cells which are made gradually move towards the surface of the skin. As they go, they die, and fill up with a protein called **keratin**. The top layer of the skin is made up of these dead cells. It is called the **cornified layer**.

The cornified layer protects the softer, living cells underneath, because it is hard and waterproof. It is always being worn away, and replaced by cells from beneath. On the parts of the body which get most wear – for example, the soles of the feet – it grows thicker.

Some of the cells in the epidermis contain a dark brown pigment, called **melanin**. Melanin absorbs the harmful ultra-violet rays in sunlight, which would damage the living cells in the deeper layers of the skin.

Here and there, the epidermis is folded inwards, forming a **hair follicle**. A hair grows from each one. Hairs are made of keratin. Each hair follicle has a **sebaceous gland** opening from the side of it. These glands make an oily liquid called **sebum**. Sebum keeps the hair and skin soft and supple.

> **Question**
> Explain the meaning of the terms
> (a) homeothermic, and (b) poikilothermic.

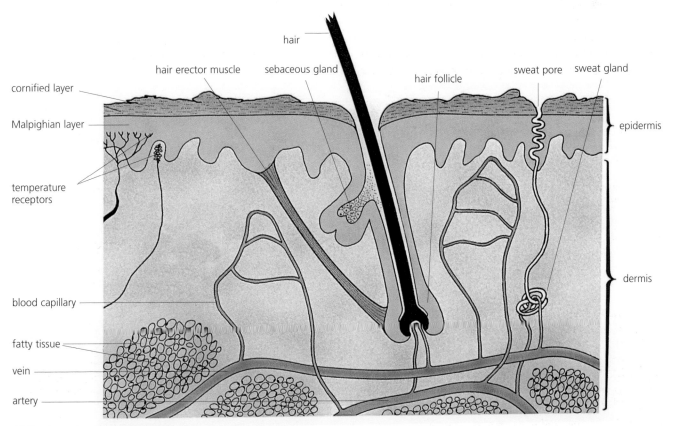

10.2 A section through human skin.

10.6 The dermis has many functions.

Most of the dermis is made of **connective tissue**. This tissue contains elastic fibres. As a person gets older, the fibres lose their elasticity, so the skin becomes loose and wrinkled.

The dermis also contains **sweat glands**. These secrete a liquid called **sweat**. Sweat is mostly water, with small amounts of salt and urea dissolved in it. It travels up the sweat ducts, and out onto the surface of the skin through the sweat pores. As we will see, sweat helps in temperature regulation.

The dermis contains **blood vessels**, and **nerve endings**. The nerve endings are sensitive to touch, pain, pressure and temperature, so they help to keep you aware of changes in your environment.

Underneath the dermis is a layer of **fat**. This is made up of cells which contain large drops of oil. This layer helps to insulate your body, and also acts as a food reserve.

10.7 The skin reacts when you are too cold.

When your body temperature drops below 37 °C, nerve impulses from the hypothalamus cause your skin to reduce the rate of heat loss (Fig 10.3). Firstly, the impulses make the **erector muscles** attached to your hair follicles contract, which makes the hairs stand on end. If you were a furry animal, this would provide you with a thick layer of hair, trapping air – which is an excellent insulator – next to your skin. As it is, you just get goose pimples, which don't help at all!

However, a second effect of the impulses from the hypothalamus is more useful in conserving heat. The impulses make the muscles around the **arterioles**, which supply blood to the capillaries near the surface of your skin, contract. This closes off these arterioles, and stops blood flowing along this pathway. This is called **vasoconstriction**. The blood now has to go through the capillaries which lie below the fat layer. The fatty tissue is a good insulator, so this reduces the amount of heat which is lost by radiation from your blood to the air. Your skin feels cold – but your blood will stay warm.

There are other responses, which do not involve your skin, which happen when your blood is too cold. You will probably **shiver**. Shivering is a very fast, random contraction and relaxation of muscles, which generates heat to warm your blood. Metabolic reactions in the liver may speed up, again generating extra heat. You may do something active like jumping up and down, which also increases the amount of heat generated in your body. You can also do things to decrease the amount of heat lost by your body, such as moving into a warmer place, or putting on more clothes.

10.8 The skin reacts when you are too warm.

If your blood temperature rises much above 37 °C, the hypothalamus sends nerve impulses which cause a different set of responses by your skin.

The erector muscles relax, so that the hairs lie flat against the skin, no longer trapping as much air and so allowing more heat to be lost by radiation.

The muscles constricting the arterioles relax, so that the blood is free to flow through the surface capillaries. This brings it close to the surface of the skin, so that it can lose more heat by radiation. This is called **vasodilation**.

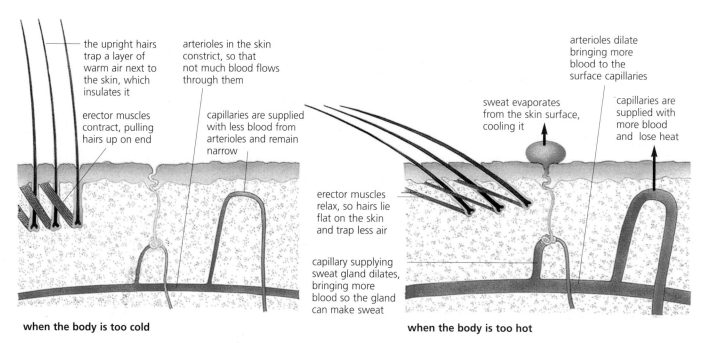

10.3 How the skin helps in temperature regulation.

The **sweat glands** begin to secrete **sweat**. This is a fluid made from blood plasma, containing mostly water and some salt and a little urea. It flows from the glands, through the sweat ducts and onto the surface of the skin. Here the water in it evaporates. The evaporation of water takes a lot of heat energy, so heat is lost from the skin. This is a very important process, because it is the only way we have of making our body temperature *lower* than that of the temperature around us.

You can also purposefully do things which help you to keep cool. You can lie still, so generating less heat within your body. You can take off layers of clothes, or wear light, loose-fitting clothes. You can swim; when you get out of the water, the water evaporates from your skin and cools you down. You can move into the shade or a cool room.

10.9 Temperature regulation involves negative feedback.

Fig 10.4 summarises the way in which the hypothalamus, skin and muscles work together to keep your body temperature constant.

We have seen that, when the temperature of your blood rises above the norm, the hypothalamus senses this. It responds by sending nerve impulses to your skin that bring about actions to help cool the blood. When the cooler blood reaches the hypothalamus, it responds by sending nerve impulses to your skin that bring about actions to help reduce the rate at which heat is lost from the blood. At the same time, the rate of heat production in the muscles is increased.

So, all the time, the hypothalamus is monitoring small changes in the temperature of your blood. As soon as it rises above normal, actions take place that help to reduce the temperature. Then, as soon as the hypothalamus senses the lowered temperature, it stops these actions taking place and starts off another set of actions that help to raise the blood temperature.

This process is called **negative feedback**. The term 'feedback' refers to the fact that, when the hypothalamus has made your skin take action to increase heat loss, information about the effects of these actions is 'fed back' to it, as it senses the drop in the blood temperature. It is called 'negative' because the information that the bood has cooled stops the hypothalamus making your skin do these things.

> ### Practical 10.1 Investigating the effect of size and covering on rate of cooling
>
> 1 Draw a results table.
> 2 Heat some water to about 80°C.
> 3 While the water is heating, label three test tubes as follows: small tube A and two equal-sized large tubes B and C.
> 4 Wrap dry cotton wool around tube C, and hold in place with a rubber band.
> 5 When the water has reached 80°C, pour it into the three tubes, to the same level in each one. *Immediately* take the temperature in each tube, and then put a cork into each tube as quickly as you can. Record this initial temperature in your results table.
> 6 Take the temperature of the water in each tube every 5 minutes for the next 30 minutes. Each time, replace the corks in the tubes as quickly as possible.
> 7 Plot your results on graph paper, with time on the horizontal axis and temperature on the vertical one. Draw a separate best fit line for each tube.
>
> ### Questions
> 1 Which tube cooled fastest?
> 2 Did tube A cool faster or more slowly than tube B? Explain why.
> 3 Did tube B cool faster or more slowly than tube C? Explain why.
> 4 In cold climates, animals lose heat from their bodies just as the test tubes did. Would you expect a small animal to lose heat faster or more slowly than a large animal?
> 5 What do mammals have on their skin to help them to conserve heat?
> 6 Polar bears are the biggest of all the bears, whereas Malayan sun bears, which live in hot countries, are much smaller. Can you explain how this might help the bears survive in their different environments, using the results from this experiment?

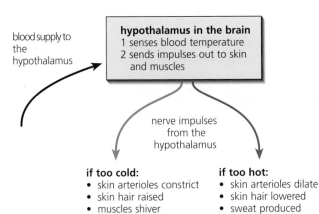

10.4 Regulation of body temperature.

Practical 10.2 Investigating the effect of evaporation on rate of cooling

When a liquid changes to a gas, it is said to **evaporate**. Evaporation uses heat energy, so when a liquid on your skin evaporates it takes heat energy from your skin, cooling it down.

1. Set up two stands with clamps and bosses, ready to hold thermometers. You will also need two pieces of cotton wool and two rubber bands. Soak one of the pieces of cotton wool in cold water.
2. Collect two thermometers, which have been kept in a warm place. As quickly as possible, wrap dry cotton wool round the bulb of one of them, and wet cotton wool around the other. Support each thermometer in a clamp, and put them side by side on the bench in front of you.
3. Every two minutes, read the temperature on each thermometer, recording it in a results table. Do this for 20 minutes.
4. Draw graphs to show how each thermometer cooled.

Questions

1. Why were the two thermometers supported in clamps, rather than held in your hand or left lying on the bench?
2. Which thermometer cooled fastest? Explain why you think this was so.
3. If someone is rescued after falling into cold water, it is very important to get them into dry clothing as soon as possible. Why?
4. How do the following living organisms use evaporation to help them to cool their bodies: (a) a human (b) a dog (c) a tree?
5. Some of the hottest places in the world are dry deserts, such as the Sahara. What special problems would mammals have in using evaporation to keep cool if they lived in a desert?

Question

Polar bears live in very cold climates.
Using your general knowledge and reference books explain what special adaptations polar bears have to help them to keep warm.

The control of blood glucose content

10.10 Glucose is needed for respiration.

All cells in your body respire all the time. They must do this to provide themselves with a supply of energy. If they stop respiring, they die.

For many of your cells, the fuel they use in respiration is **glucose**. It is therefore very important that the concentration of glucose in your blood and tissue fluid is kept fairly constant, so your respiring cells do not run out. This is especially important for brain cells, which quickly die if they become short of glucose. It is also dangerous to have too much glucose in your blood. If the blood becomes more concentrated than the cytoplasm in your cells, then water will leave the cells by osmosis. The cells may die.

Glucose concentration in the blood is normally around 100 mg of glucose in every 100 cm^3 of blood. This is controlled by the **pancreas** and the **liver**.

10.11 The pancreas secretes insulin and glucagon.

The pancreas is a soft, white structure which lies just below your stomach, on the left-hand side of the body below the ribs. It is an unusual organ because it has two very different functions. Most of the pancreas is made up of cells which secrete pancreatic juice, which flows along the pancreatic duct into the duodenum, where it helps in digestion (Section 4.38). But dotted around in the pancreas are groups of a different kind of cell; the groups are called **islets of Langerhans**. These cells secrete the hormones **insulin** and **glucagon**.

The cells in the islets of Langerhans constantly monitor the amount of glucose in the blood. If this level rises too high, then they secrete insulin. If the blood glucose level falls too low, they secrete glucagon.

10.12 The liver changes the amount of glucose in the blood.

Insulin and glucagon are carried around the body, in solution in blood plasma. Both of them affect the behaviour of the liver cells (Fig 10.5).

Insulin makes the liver change glucose into the polysaccharide **glycogen,** which it stores inside its cells. Insulin also increases the rate at which the liver cells use up glucose in respiration. This reduces the amount of glucose in the blood.

Glucagon makes the liver release glucose into the blood, so increasing the blood glucose level. The glucose that the liver releases is made by breaking apart the glycogen molecules stored in its cells.

Note – If you are studying 'O' level, you only need to know about insulin, not glucagon.

10.13 Lack of insulin causes sugar diabetes.

Some people's pancreases do not produce enough insulin. If this happens, it is difficult to control the amount of glucose in the blood. This disease, called **diabetes mellitus**, can be very dangerous.

Imagine a person has eaten a meal containing a lot of sugar. Normally, the pancreas will respond by producing insulin, which will reduce the amount of glucose in the blood. But if the person is diabetic, almost all the glucose will go into the bloodstream and be taken round the body.

This is dangerous. Very high blood glucose levels can damage brain cells, causing coma and possibly death.

The extra glucose in the blood is excreted by the kidneys. One way of diagnosing diabetes is to test urine for glucose. Healthy people have no glucose at all in their urine.

If there is no insulin in the body, then the liver and muscles will not build up stores of glycogen. So later, when the blood glucose levels have dropped again, there will be no reserves to draw on. The blood glucose may drop to a very low level, again possibly causing coma and death.

In another form of diabetes mellitus, the pancreas does still make insulin, but the liver cells do not respond to it. The symptoms are just the same as if no insulin is made at all.

Neither form of diabetes can be cured, but they can be controlled. Most people with diabetes can remain healthy by eating small amounts of sugary or starchy foods at regular intervals. Some, however, need to inject themselves with insulin each day.

Questions

1. Why do body cells need glucose?
2. In healthy humans, the blood contains 60 – 110 mg of glucose per 100 cm^3 of blood. Which gland secretes the hormones which are responsible for keeping this level fairly constant?
3. The graph in Fig 10.6 shows the changes in blood glucose level after a meal.

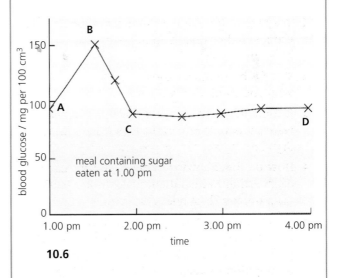

10.6

Explain the shape of the graph between (a) A and B, (b) B and C, and (c) C and D.

4. People with one form of the disease diabetes mellitus cannot make insulin. Why is it dangerous for diabetics to eat a meal containing a lot of sugar?
5. The regulation of blood glucose levels is one example of homeostasis. What is homeostasis, and why is it important?

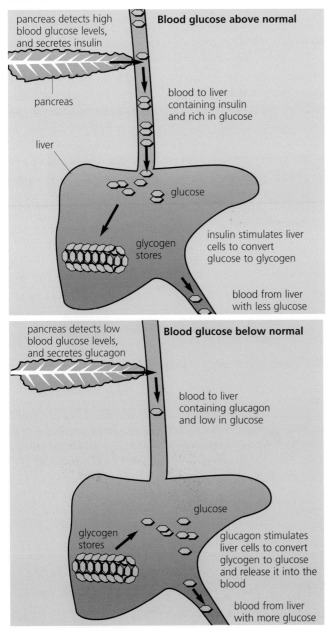

10.5 How blood sugar is regulated.

Excretion

10.14 Excretion is the removal of waste substances.

The metabolic reactions taking place in cells constantly produce waste products, such as carbon dioxide. Some of these waste products can damage the cells – they are **toxic**. Such toxic substances must be removed from the body. The removal of toxic waste materials and products of metabolism is called **excretion**.

Humans have three main excretory substances (Fig 10.7):

- **Carbon dioxide**
  This is made by all cells, in respiration. It is transported in blood plasma to the alveoli of the lungs, where it is excreted in the air you breathe out.
- **Bile pigments**
  These are made in the liver, from the haemoglobin in old red blood cells. They are carried in bile into the duodenum, and excreted in the faeces.
- **Urea**
  This is made in the liver, from excess proteins. It is carried in blood plasma to the kidneys, where it is excreted in urine. It contains nitrogen, so it is a nitrogenous excretory product.

10.15 Excess proteins are converted to urea.

When you eat proteins, digestive enzymes in your stomach, duodenum and ileum break them down into amino acids. The amino acids are absorbed into the blood capillaries in the villi in your ileum (Section 4.41). The blood capillaries all join up to the hepatic portal vein, which takes the absorbed food to the liver.

The liver allows some of the amino acids to carry on, in the blood, to other parts of your body. But if you have eaten more than you need, then some of them must be changed into something else.

It would be very wasteful to excrete the extra amino acids just as they are. They contain energy which, if it is not needed straight away, might be needed later.

So enzymes in the liver split up each amino acid molecule (Figs 10.8 and 10.9). The part containing the energy is kept, turned into carbohydrate or fat, and stored. The rest, which is the part which contains nitrogen, is combined with carbon dioxide and turned into urea. This process is called **deamination**.

The urea dissolves in the blood plasma, and is taken to the kidneys to be excreted. A small amount is also excreted in sweat.

The liver has many other functions, as well as deamination. Some of these are listed in Table 10.1.

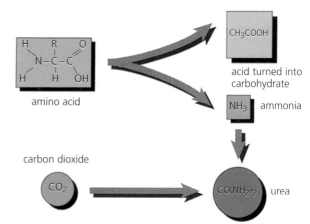

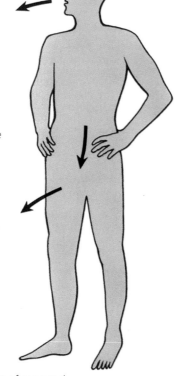

10.8 Deamination and urea formation.

10.7 Excretory products of mammals.

Table 10.1 Some functions of the liver.

1 Converts excess amino acids into urea and carbohydrate, in a process called deamination
2 Controls the amount of glucose in the blood, with the aid of the hormones insulin and glucagon
3 Stores carbohydrate as the polysaccharide glycogen
4 Makes bile
5 Breaks down old red blood cells, storing the iron, and excreting the remains of the pigments in bile
6 Breaks down hormones and harmful substances such as alcohol
7 Stores vitamins D and A
8 Makes cholesterol, which is needed to make and repair cell membranes

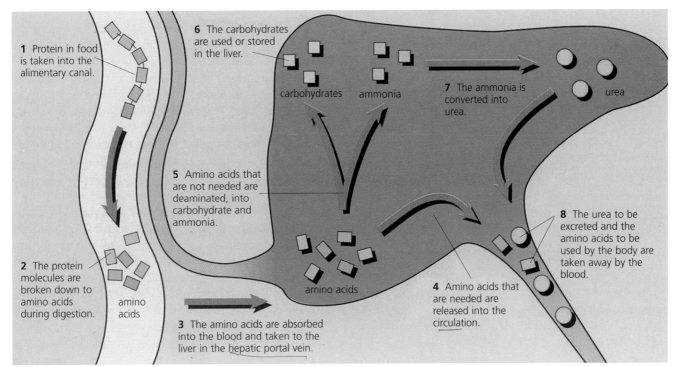

10.9 How urea is made.

> **Questions**
> 1 What is meant by an excretory product?
> 2 What are the two main excretory products of plants?
> 3 Through which part of the plant are these substances excreted?
> 4 Why is urea called a nitrogenous excretory product?
> 5 Why do plants not have nitrogenous excretory products?
> 6 What happens to excess amino acids in the liver?

The human excretory system

10.16 The kidneys are part of the excretory system.

Fig 10.10 illustrates the position of the two kidneys in the human body. They are near the back of the abdomen, behind the intestines.

Leading from each kidney is a tube, called the **ureter**. The ureter carries urine that the kidney has made to the **bladder**.

10.17 Urine is made by filtration and reabsorption.

The function of the kidneys is to take unwanted substances from the blood and to pass them on to the bladder, to be excreted.

Blood is brought to the kidneys in a branch of the renal artery. The blood is filtered into tiny tubules inside the kidneys. All of the glucose, and some of the salts, that have gone into the tubules are taken back, or **reabsorbed**, into the blood. The remaining fluid in the tubules is water with urea and some salts dissolved in it. This fluid is called **urine**.

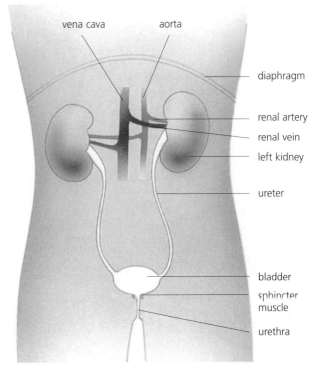

10.10 The human excretory system.

10.18 The bladder stores urine.

The urine from both kidneys flows into the ureters. The ureters take it to the bladder.

The bladder stores urine. It has stretchy walls, so that it can hold quite large quantities.

Leading out of the bladder is a tube called the **urethra**. There is a sphincter muscle at the top of the urethra, which is usually tightly closed. When the bladder is full, the sphincter muscle opens, so that urine flows along the urethra and out of the body.

Adult mammals can consciously control this sphincter muscle. In young mammals, it opens automatically when the bladder gets full.

10.19 Kidney machines can do the work of damaged kidneys.

Sometimes, a person's kidneys may stop working properly. This might be because of an infection in the kidneys. Complete failure of the kidneys allows urea and other waste products to build up in the blood. It also means that the amount of water in the body is not regulated. This will cause death unless treated.

If a transplant is not possible, then people with kidney failure are treated regularly on a **kidney machine**. Fig 10.11 shows one kind of kidney machine.

Blood from an artery in the patient's arm flows into the kidney machine. Here, it passes through a **dialyser**. The dialyser contains fluid made up of water and other substances, such as salt.

As the patient's blood passes through the dialyser, it is separated from the fluid by a partially permeable membrane. As there is no urea in the fluid, most of the urea in the blood diffuses through the membrane into the fluid. The amount of other substances remaining in the blood can be regulated by controlling their concentrations in the fluid. The blood cells, and any large molecules such as proteins, remain in the blood as they are too big to pass through the holes in the dialysis membrane.

Patients have to be treated on a kidney machine two or three times a week. Each treatment lasts several hours.

Although treatment on a dialyser will keep a patient alive and well, it is not the ideal solution. Some people feel unwell during dialysis. It is also very inconvenient. Some patients have a dialyser at home, but many have to go into hospital for treatment, because kidney machines are expensive so there are not enough for all patients to have their own.

A better long-term solution for kidney failure is a kidney transplant. However, this has drawbacks as well. There are not enough donor kidneys available, and they are often rejected by the recipients even if the transplant operation is successful. You can read more about the immune system in Chapter 13.

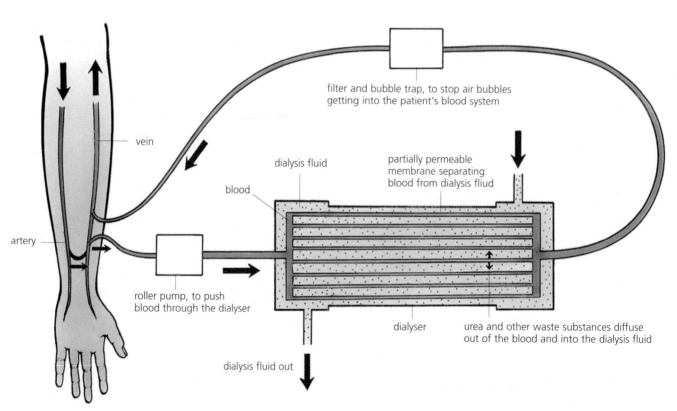

10.11 How kidney dialysis works.

Questions

1. Which blood vessels bring blood to the kidneys?
2. What is urine?
3. Where are (a) ureters, and (b) the urethra found?
4. Look at Fig 10.11.
   (a) Why is the blood passed through the dialyser in many small channels instead of one large one?
   (b) Why might it be dangerous if an air bubble gets into the patient's blood?
   (c) Give two reasons why a successful kidney transplant is better than treatment on a kidney machine.

Chapter revision questions

1. (a) What is the main nitrogenous waste product excreted by human kidneys?
   (b) Where is this waste product formed?
   (c) Briefly describe how this waste product is formed.
   (d) Which blood vessels deliver this waste product to the kidneys?
   (e) Name two substances found in the blood plasma which are not found in the urine of a healthy person.

2. A student carried out an experiment to compare the rates at which water cooled in two containers, A and B. Container A was an ordinary glass beaker. Container B was a similar beaker which had been wrapped in a dry, woolly material. Both containers, and a vacuum flask, were filled with hot water from the same supply and each was covered with a lid. At the beginning of the experiment, the temperature of the water in all three containers was 86 °C. The temperature in the laboratory was kept constant at 18 °C. The student recorded the temperature of the water in A and B alternately every two minutes for twenty minutes. The results are shown in the table.

Time / min	Temperature / °C		
	A	B	vacuum flask
0	86	86	86
2		80	
4	66		
6		70	
8	50		
10		61	
12	50		
14		56	
16	32		
18		53	
20	28	52	86

(a) On graph paper, plot the readings for container A and for container B so that they can be distinguished.
Draw and label the curves for A and B, but do not include the reading taken at 12 minutes in the curve for A. [4]
(b) (i) Suggest why the temperature at 12 minutes should **not** be part of the curve. [1]
(ii) Suggest how this reading might have been obtained. [1]
(iii) By referring to your graph, suggest what the reading from container A for 12 minutes should have been. [1]
(c) In a mammal the structure of the skin can affect the rate at which heat is lost from the body. Explain how this is shown by the experiment. [2]
(d) State three ways in which the skin can help to keep the body temperature constant if the external temperature falls. [3]
5090/6 N00

3. (a) Explain how, in a healthy person, (i) the body temperature and (ii) the blood sugar, are returned to normal after they have risen above normal levels. [10]
(b) Explain why it is important for humans to maintain a constant body temperature of 37 °C.
5090/2 J00 [2]

4. (a) Distinguish between excretion and egestion. [4]
(b) Describe the passage of water from blood in the aorta to its excretion via the urethra. Illustrate your answer with the aid of a simple, labelled diagram. [7]
(c) Outline the role of the liver in excretion. [4]

0610/3 J00

11 Support and movement

The human skeleton

11.1 Bone is made of protein and minerals.

All living organisms are held in shape, or supported, in some way. Many of them have special structures which do this. These structures are called **skeletons**. The human skeleton is made of bone and cartilage.

Most of the human skeleton (Fig 11.1) is made of **bone**. Bone is mostly made of mineral substances such as calcium phosphate, with small amounts of magnesium salts. This makes it very hard. Bone also contains stretchy fibres of a protein called **collagen**, which give it elasticity.

Bone is alive. It contains living cells, which are supplied with food and oxygen by blood vessels.

11.2 Cartilage contains fewer minerals than bone.

Cartilage is much softer than bone. This is because it does not contain very many mineral salts, but like bone, it contains collagen.

Cartilage is found on the ends of bones, where they meet one another at a joint. It allows the bones to move easily over each other because it is smooth. There is also cartilage in the pinnae of your ears, and in the end of your nose.

11.3 Bones are joined in different ways.

Wherever two bones meet each other, a **joint** is formed. Sometimes two bones are joined quite firmly together by fibres. The bones in the cranium of the skull are joined like this. The bones are held so tightly together in an adult human that they cannot move at all.

Table 11.1 Functions of the human skeleton.

Function	Example
Support	Vertebral column Pectoral girdle Pelvic girdle Leg bones
Movement	Leg and arm bones Vertebral column
Protection	Skull (protects brain) Ribs (protect heart and lungs) Vertebral column (protects spinal cord)
Making red and white blood cells	Marrow in leg bones and ribs

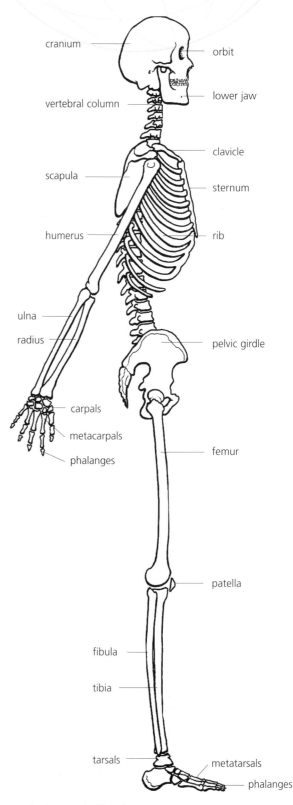

11.1 The human half-skeleton.

Other joints, however, allow the two bones to move relative to each other. They are called **synovial joints**.

The elbow joint and shoulder joint (Fig 11.2) are examples of synovial joints. Fig 11.3 shows the structure of a typical synovial joint. The two bones are held together by **ligaments**. Ligaments are very strong, but can stretch when the bones move.

If the two bones rubbed against one another when they moved, they would quickly be damaged. So the ends of the bones are covered with a layer of smooth, slippery cartilage. Between the bones is a small amount of a thick liquid called **synovial fluid**. This lubricates the joint, so that it moves smoothly. The fluid is made and kept in place by the synovial membrane.

Synovial joints are given different names, depending on the kind of movement that takes place. The elbow joint is a **hinge joint**, because the bones can only move in one plane, like a door on hinges. The shoulder and hip joints are **ball and socket joints**. A ball at the end of one bone fits into a socket in the other. This allows a circular movement, or movement in all planes.

11.4 Muscles can contract.

Muscles are made of a very special kind of tissue. The cells in the tissue can **contract**; that is, they can make themselves shorter. They use quite a lot of energy to do this, which they get by breaking down glucose in respiration.

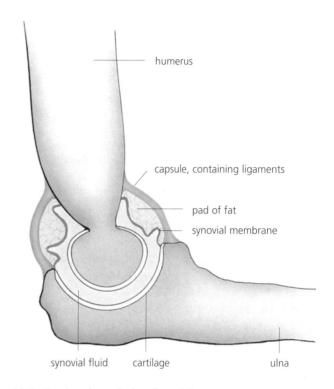

11.3 Section through the elbow joint.

11.4 Muscle cells.

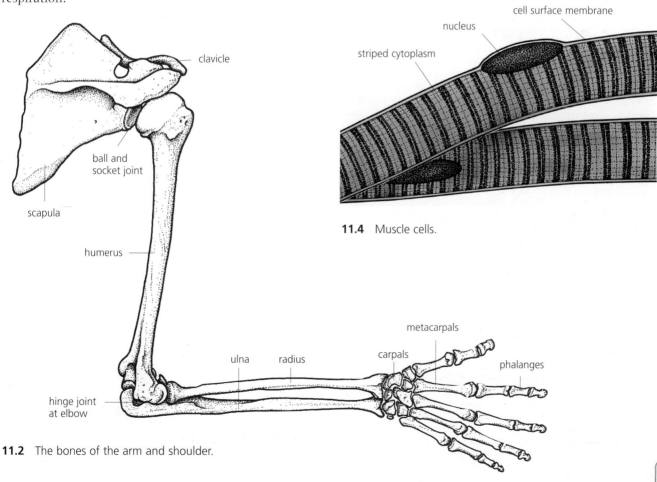

11.2 The bones of the arm and shoulder.

Questions

1. What are the two main constituents of bone?
2. Give one similarity and one difference between bone and cartilage.
3. What is meant by a synovial joint?

It is important to remember that muscles can only contract and relax. When they contract, they get shorter. When they relax, they stay just how they are – they cannot make themselves get longer. The only way a muscle can be made longer is if something pulls on it. We will see how this works when we look at how the muscles in your arm cause movement at the elbow joint.

11.5 Two muscles move the forearm.

Fig 11.5 shows the bones and two of the muscles in your arm. The arm can bend at the elbow, which is a hinge joint.

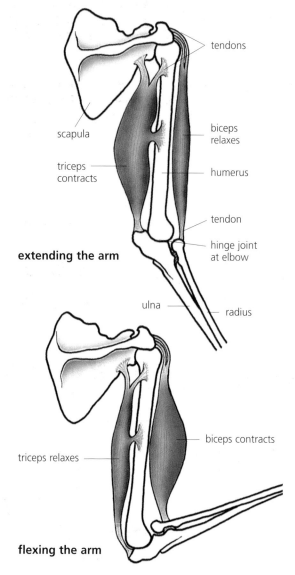

11.5 Movement of the forearm.

The **biceps muscle** is attached to the scapula at the top, and the radius at the bottom. It is attached to the bones by strong **tendons**. When it contracts, it pulls on the tendons and this pulls the radius and ulna up towards the scapula, so the arm bends. This is called **flexing** your arm, so the biceps is a **flexor muscle**.

But muscles can only pull, not push. The biceps cannot push your arm back down again. Another muscle is needed to pull it down. The **triceps muscle** does this. When it contracts, the triceps straightens or extends your arm. It is called an **extensor muscle**. A pair of muscles working together like this are called **antagonistic** muscles.

Support in plants

11.6 Xylem forms wood and supports plants.

Plant stems, roots and leaves contain xylem (Section 7.23). Xylem is made of cells which have very strong walls containing lignin. These lignified xylem vessels help to support the stem.

The larger and taller the plant, the more support it needs. Trees are supported by the wood in their trunks and branches, which is made almost entirely of xylem.

11.7 Cell turgor supports herbaceous plants.

In parts of plants where there is not much xylem, another means of support is needed.

When a plant has plenty of water, the contents of each cell press outwards on the cell wall. This makes the cells firm, or **turgid**. They press against each other, holding the plant firm and upright. This is particularly important in leaves, and in **herbaceous plants**, which do not have woody stems.

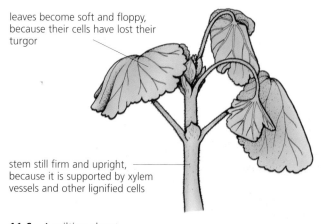

11.6 A wilting shoot.

Questions

1. What is wood made of?
2. How does turgor help to support plants?

12 Inheritance and evolution

12.1 Nuclei contain chromosomes carrying genes.

In the nucleus of every cell there are a number of **chromosomes**. Chromosomes are long threads made of DNA and protein.

Most of the time, the chromosomes are too thin to be seen except with an electron microscope. But when a cell is dividing, they get shorter and fatter (Figs 12.1, 12.2 and 12.3), so they can be seen with a light microscope.

Each chromosome contains one very long molecule of DNA. The DNA molecule carries a code that instructs the cell about which kinds of proteins it should make. Each chromosome carries instructions for making many different proteins. A part of a DNA molecule coding for one protein is called a **gene**.

It is the genes on the chromosomes which determine all sorts of things about you – what colour your eyes or hair are, whether you have a snub nose or a straight one, and whether you have a genetic disease such as cystic fibrosis.

12.2 Each species has its own set of genes.

Each species of organism has its own number and variety of genes. This is what makes their body chemistry, their appearance and their behaviour different from those of other organisms.

Humans have a large number of genes. You have 46 chromosomes inside each of your cells, all with many genes on them. Every cell in your body has an exact copy of all your genes. But, unless you are an identical twin, there is no-one else in the world with exactly the same combination of genes that you have. Your genes make you unique.

We are now able to take a gene out of a cell of one organism, and place it in another. This is called **genetic engineering**, and you can read more about it in Chapter 14.

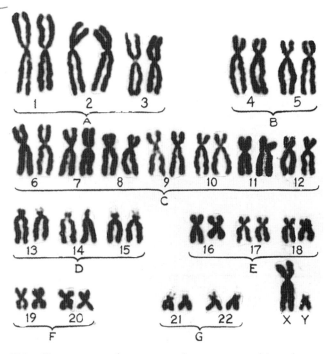

12.2 Chromosomes from a normal man, arranged in order.

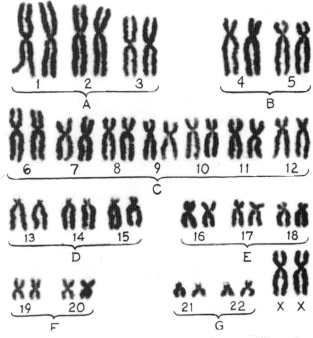

12.3 Chromosomes from a normal woman, arranged in order.

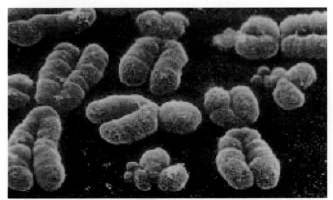

12.1 A scanning electron micrograph of human chromosomes. You can see that each one is made of two chromatids, linked at a point called the centromere.

119

12.3 Most cells contain pairs of homologous chromosomes.

Fig 12.1 is a photograph of some chromosomes from a human cell which is about to divide. In Figs 12.2 and 12.3 photographs of all the chromosomes have been rearranged. You can see that there are, in fact, 23 pairs of chromosomes. The two chromosomes in a pair are the same size and shape.

The two chromosomes of a pair are called **homologous chromosomes**. One came from the person's mother, and the other from the father.

Each chromosome of a homologous pair carries genes for the same characteristic in the same place (Fig 12.4). Because there are two of each kind of chromosome, each cell contains two of each kind of gene. Let us look at one kind of gene to see how it behaves, and how it is inherited.

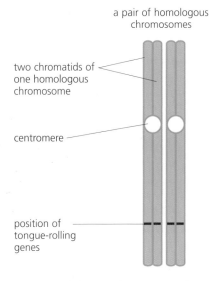

12.4 Homologous chromosomes have genes for the same characteristic in the same position.

12.4 Cystic fibrosis is caused by an unusual protein.

In humans, cells in the lungs make mucus. The kind of mucus which is made partly depends on a **protein** in the cell membrane of some of the cells in the lungs. In some people, this protein is made incorrectly. This causes the disease **cystic fibrosis**. The lungs of a person with this disease make too much mucus, which is thicker than usual. The mucus collects in the lungs, making it difficult to get enough oxygen into the blood (Fig 12.5). The mucus makes a good breeding ground for bacteria, so infections can build up. The thick mucus is also made in the pancreas, where it blocks the pancreatic duct. This prevents pancreatic juice, containing digestive enzymes, from flowing into the duodenum, so food cannot be digested properly.

Instructions for making the protein are given by a gene. There are two varieties of the gene for this protein – one for making the normal protein, and one for making the incorrect one. Different varieties of a gene are called **alleles**. We can give letters to alleles to use as symbols. So we can call the normal allele of the gene F, and the one which codes for the abnormal protein f.

12.5 People with cystic fibrosis have to have frequent therapy to clear mucus from their lungs. This boy's father pummels his back to dislodge the thick mucus, which the boy can then remove from his lungs.

12.5 Each cell has two genes for any characteristic.

In each of your cells, there are two genes giving instructions about which of these two kinds of protein to make. This means that there are three possible combinations of alleles. You might have two F alleles, FF. You might have one of each, Ff. Or you might have two f alleles, ff (Fig 12.6).

If the two alleles for this gene in your cells are the same, that is FF or ff, you are said to be **homozygous**. If the two alleles are different, that is Ff, then you are **heterozygous**.

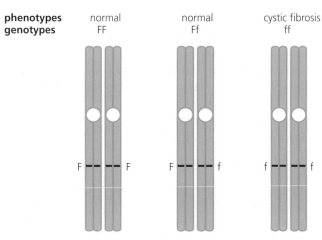

12.6 Genotypes for the cystic fibrosis gene.

12.6 Genotype can determine phenotype.

The genes that you have are your **genotype**. Your genotype could be FF, Ff or ff.

The genotype determines the kind of protein you have, and therefore whether you will have cystic fibrosis or not. If your genotype is FF, then your protein is normal, and you will not have the disease. If your genotype is ff, then your protein is the incorrect one, and you will have cystic fibrosis. If your genotype is Ff, some of your protein is the incorrect one, and some of it is normal. You will have enough normal protein to ensure that the right sort of mucus is made, and you will not suffer from cystic fibrosis.

The features you have are called your **phenotype**. This can include what you look like – for example, what colour your hair is, or how tall you are – as well as things which we cannot actually see, such as what kind of protein you have in your cell membranes. In our example, your phenotype is either being normal or having cystic fibrosis.

You can see that, in this example, your phenotype depends entirely on your genotype. This is not always true. Sometimes, other things, such as what you eat, can affect your phenotype. However, for the moment, we will only consider the effect which genotype has on phenotype, and not worry about effects which the environment might have.

12.7 Alleles can be dominant or recessive.

You have seen that there are three different possible genotypes for the cell surface membrane protein, but only two phenotypes. We can summarise this as follows:

genotype	phenotype
FF	normal
Ff	normal
ff	cystic fibrosis

This happens because the allele F is **dominant** to the allele f. A dominant allele has just as much effect on phenotype when there is one of it as when there are two of it. A person who is homozygous for a dominant allele has the same phenotype as a person who is heterozygous. A heterozygous person is said to be a **carrier** of cystic fibrosis, because he or she has the allele for it but shows no symptoms.

The allele f is **recessive**. A recessive allele only affects the phenotype when there is no dominant allele present. Only people with the genotype ff – homozygous recessive – have cystic fibrosis.

12.8 Some alleles show codominance.

Sometimes, neither of a pair of alleles is dominant or recessive. Instead, both of them have an effect on the phenotype of a heterozygous organism. This is called **codominance**.

You will have noticed that, when we were dealing with dominant and recessive alleles, we used a capital letter for the dominant allele and a small letter for the recessive allele. We cannot do that with codominant alleles, because it might give the impression that one was dominant and the other recessive. Instead, we use a capital letter to represent the gene, and then small letters written just above and to the right of it to represent the alleles of that gene.

For example, imagine a kind of flower which has two alleles for flower colour (Fig 12.7). The allele C^W produces white flowers, while the allele C^R produces red ones. The possible genotypes and phenotypes are:

genotype	phenotype
$C^W C^W$	white
$C^W C^R$	pink
$C^R C^R$	red

Most people do not like writing alleles like this – it is much easier to try to get away with just writing a single letter! But it is very important that you *do* use the correct kind of symbols, or people will not understand that the alleles are codominant.

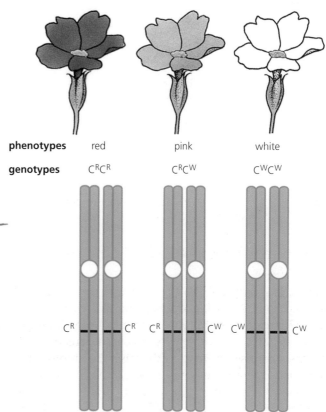

12.7 Codominance.

Questions

1. What are chromosomes made of?
2. Explain how genes affect all the chemical reactions in an organism's body.

Another example of codominant alleles occurs in the gene that codes for blood group. This gene is given the symbol I. It has three alleles, I^A, I^B and I^O.

Alleles I^A and I^B are codominant. So the possible genotypes and phenotypes involving these two alleles are:

$I^A I^A$ blood group A
$I^B I^B$ blood group B
$I^A I^B$ blood group AB

Allele I^O is recessive to both of the other alleles. So the other possibilities for a person's blood group are:

$I^A I^O$ blood group A
$I^B I^O$ blood group B
$I^O I^O$ blood group O

Inheritance

12.9 Gametes have only one gene for any characteristic.

All of the cells in your body have come from one original cell. This first cell, the zygote, was made when a sperm fertilised an egg. The sperm and the egg each contained chromosomes with a certain set of genes on them, which have since been copied exactly into all your body cells.

Gametes, such as eggs and sperm, are made by meiosis. During meiosis, the chromosomes come together in their homologous pairs, and then separate from each other. Each daughter cell gets only one of each pair of chromosomes. Cells like this are called **haploid** cells.

So, whereas normal, **diploid**, body cells have two of each kind of chromosome, gametes only have one of each kind. This also means that they only have one of each pair of alleles.

Fig 12.8 shows some of the stages in meiosis, to show what happens to one pair of genes when a cell in the testis divides to produce sperm cells.

In this example, the person is a carrier of cystic fibrosis, genotype Ff. For simplicity, only the chromosomes carrying this gene are shown. The other 44 have been left out.

At the end of meiosis there are four cells, which will all grow into sperm cells. Half of them have the F allele and half have the f allele.

12.10 Genes and fertilisation.

If this man's partner is a woman with the genotype ff, will their children have cystic fibrosis or not?

The eggs that are made in the woman's ovaries are also made by meiosis. If you use Fig 12.8 to see what happens to her chromosomes during meiosis, you will see that she can only make one kind of egg. All of the eggs will carry an f allele.

During sexual intercourse, hundreds of thousands of sperm will begin a journey towards the egg. About half of them will carry an F allele, and half will carry an f allele. If there is an egg in the woman's oviduct, it will probably be fertilised. There is an equal chance of either kind of sperm getting there first.

If a sperm carrying an F allele wins the race, then the zygote will have an F allele from its father and an f allele from its mother. Its genotype will be Ff. After nine months, a baby will be born with the genotype Ff.

But if a sperm carrying an f allele manages to fertilise the egg, then the baby will have the genotype ff, like its mother (Fig 12.9).

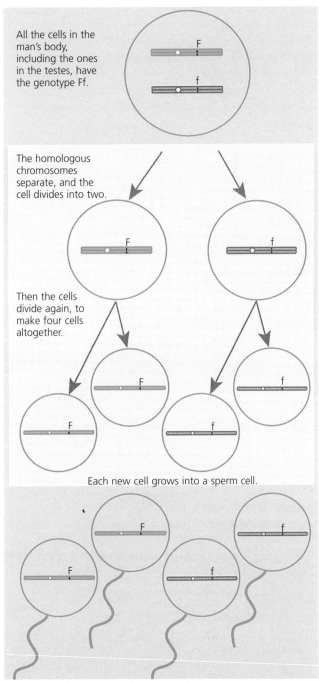

12.8 What happens to the genes of a heterozygous man during meiosis.

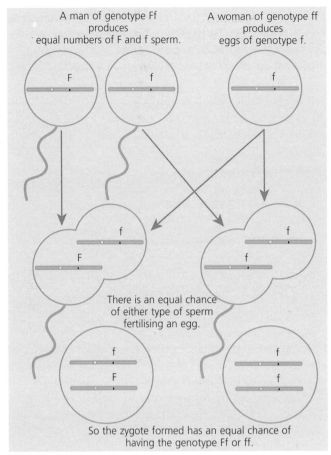

12.9 Fertilisation between a heterozygous man and a woman with cystic fibrosis.

12.11 Genetic crosses must be written clearly.

There is a standard way of writing out all of this information. It is called a **genetic diagram**. First, write down the phenotypes and genotypes of the parents. Next, write down the different types of gametes they can make, like this.

parents' phenotypes	normal	cystic fibrosis
parents' genotypes	Ff	ff
gametes	(F) or (f)	(f)

The next step is to write down what might happen during fertilisation. Either kind of sperm might fuse with an egg.

gametes	(F) or (f)	(f)
offspring genotypes	*gametes* (f)	

	(F)	Ff
	(f)	ff

offspring phenotypes Ff = normal (carrier)   50%
ff = cystic fibrosis   50%

So, we would expect that approximately half of the children would be heterozygous carriers of cystic fibrosis, and half would be homozygous, with cystic fibrosis. Another way of putting it is to say that the expected ratios of heterozygous carriers to people with cystic fibrosis would be 1:1.

12.12 Another example.

What happens if both parents are carriers?

parents' phenotypes	normal (carrier)	normal (carrier)
parents' genotypes	Ff	Ff
gametes	(F) or (f)	(F) or (f)
offspring genotypes	*gametes* (F) (f)	

	(F)	FF	Ff
	(f)	Ff	ff

offspring phenotypes FF = normal   25%
Ff = normal (carrier)   50%
ff = cystic fibrosis   25%

About one quarter of the children would be expected to have cystic fibrosis, and three quarters to be normal. However, half of the children would be expected to be carriers.

12.13 Offspring ratios are only probabilities.

In the last example, there were four offspring at the end of the cross. This does not mean that the man and woman will have four children. It simply means that each time they have a child, these are the possible genotypes that the child might have.

When they have a child, there is a 1 in 4 chance that its genotype will be FF, and a 1 in 4 chance that its genotype will be ff. There is a 2 in 4, or rather 1 in 2, chance that its genotype will be Ff.

However, as you know, probabilities do not always work out. If you toss a coin up four times you might expect it to turn up heads twice and tails twice. But does it always do this? Try it and see.

With small numbers like this, probabilities do not always match reality. If you had the patience to toss your coin up a few thousand times, though, you will almost certainly find that you get much more nearly equal numbers of heads and tails.

The same thing applies in genetics. The offspring genotypes which you work out are only probabilities. With small numbers, they are unlikely to work out exactly. With very large numbers of offspring from one cross, they are more likely to be accurate.

So, if the man and woman in the last example had eight children, they might expect six of them to be normal and two to have cystic fibrosis. But they should not be too surprised if they have three normal children and five with cystic fibrosis.

Genetic diagrams are used to solve genetics problems.

Here is a typical question that you might be given to answer. Follow the steps you would use to answer it.

Seeds resulting from a cross between a tall pea plant and a dwarf one produced plants all of which were tall. When these plants were allowed to self-pollinate, the resulting seeds produced 908 tall plants and 293 dwarf plants.

(a) Account for these results.
(b) What would be the results of interbreeding the dwarf plants?

Firstly, work out whether or not one allele is dominant over the other. Next, decide on symbols. Then write a list of possible genotypes and phenotypes. After that, you can begin to explain the results.

There must be two alleles, one producing tallness and the other dwarfness. The plants are all either tall or dwarf, and no medium-sized ones are mentioned. So one allele must be completely dominant over the other.

When a tall plant is crossed with a dwarf plant, all the offspring are tall. This shows that it is the tall allele which is dominant.

Therefore, let us use the symbol T for the allele for tall plants, and the symbol t for the allele for dwarf plants. The allele T is dominant over the allele t, which is recessive.

The possible genotypes and phenotypes are these.

genotype	phenotype
TT	tall
Tt	tall
tt	dwarf

The first cross is between a tall pea plant and a dwarf one. The possible genotypes of the tall pea plant are TT or Tt. The dwarf plant must have the genotype tt.

If the tall plant is heterozygous, Tt, then half of its gametes would contain the t allele. During fertilisation, zygotes would be formed with the genotype tt, which would appear dwarf. But as none of the offspring of this cross are dwarf, the tall parent must be TT.

The first cross is therefore as follows.

parents' phenotypes	tall	dwarf
parents' genotypes	TT	tt
gametes	T	t
offspring genotypes	gametes T	
	t	Tt

offspring phenotypes   Tt = tall   100%

This has explained the first sentence of the question. Next, these tall plants self-pollinate, in other words, they cross with themselves.

The offspring from the first cross are all heterozygous, Tt. When they self-pollinate, the results would be expected to be as follows.

parents' phenotypes	tall	tall
parents' genotypes	Tt	Tt
gametes	T or t	T or t

offspring genotypes

gametes	T	t
T	TT	Tt
t	Tt	tt

offspring phenotypes		
TT = tall		25%
Tt = tall		50%
tt = dwarf		25%

Approximately three quarters of the offspring would be expected to be tall and one quarter short. This fits in well with the figures of 908 tall, and 293 dwarf, which is a ratio of approximately 3 tall : 1 dwarf.

Now we can answer part (b).

The dwarf plants all have the genotype tt. All of the gametes they produce will carry a t allele. Therefore all of their offspring will also be tt; that is, homozygous dwarf plants.

12.15 Test crosses help to determine genotype.

In the last example, there were two kinds of pea plant, tall and dwarf. Once you have decided that the allele for dwarfness, t, is recessive, then you know that the genotype of any dwarf plants must be tt. But a tall plant could be either Tt or TT.

If you had a tall pea plant, and wanted to know its genotype, how could you find out? The best way would be to cross the plant with a dwarf plant, and then see what kind of offspring you got from the cross. If your original plant was homozygous, TT, then all the offspring would have the genotype Tt, and be tall (see the first cross in Section 12.14).

But if your plant was heterozygous, Tt, then not all the offspring will be tall. Try working it out. You will find that half of the offspring would have the genotype tt, and be dwarf.

By crossing a plant showing the dominant characteristic (in this case, tallness) with a homozygous recessive plant (in this case, a dwarf one) you can find out the genotype of the plant with the dominant characteristic. This cross is called a **test cross**.

12.16 F1 and F2 generations.

In the cross between the pea plants in Section 12.14, the original parents were both homozygous. One had the genotype TT and the other tt. The offspring from a cross like this, where one parent is homozygous for the dominant allele and the other parent is homozygous for the recessive allele, are called the **first filial generation**, or **F1 generation**. They are always, of course, heterozygous.

The offspring of a cross between the individuals in the F1 generation are called the second filial generation, or **F2 generation**. Their phenotypes show a ratio of 3:1.

12.17 Blood groups are inherited.

We saw earlier that there are three different alleles of the gene that determines your blood group. However, this makes no difference at all to the way you draw genetic diagrams to work out how blood groups are inherited.

For example, let us say that we want to know how a child can have blood group O, while her father has blood group A and her mother has blood group B.

If you look back at the list of possible genotypes and phenotypes for blood groups, in Section 12.8, you will see a person with blood group O must have the genotype $I^O I^O$.

For the child to have this genotype, she must have inherited an I^O allele from each of her parents.

Therefore the father's genotype must be $I^A I^O$ and the mother's genotype must be $I^B I^O$.

parents' phenotypes	group A	group B
parents' genotypes	$I^A I^O$	$I^B I^O$
gametes	I^A or I^O	I^B or I^O

offspring genotypes

	I^B	I^O
I^A	$I^A I^B$	$I^A I^O$
I^O	$I^B I^O$	$I^O I^O$

offspring phenotypes		
$I^A I^B$ = blood group AB	25%
$I^A I^O$ = blood group A	25%
$I^B I^O$ = blood group B	25%
$I^O I^O$ = blood group O	25%

> **Question**
>
> Use a genetic diagram to work out the possible blood groups of the children of a man who is homozygous for blood group A, and a woman with blood group O.

12.18 Sex is determined by X and Y chromosomes.

If you look carefully at Figs 12.2 and 12.3 you will see that the last pair of chromosomes is not the same in each case. In the first photograph, which is of a man's chromosomes, the last pair are not alike. One is much smaller than the other. In the second photograph, of a woman's chromosomes, the last pair *are* alike.

This last pair of chromosomes is responsible for determining what sex a person will be. They are called the **sex chromosomes**. A woman's chromosomes are both alike and are called X chromosomes. She has the genotype XX.

A man, though, only has one X chromosome. The other, smaller one, is a Y chromosome. He has the genotype XY (Fig 12.10).

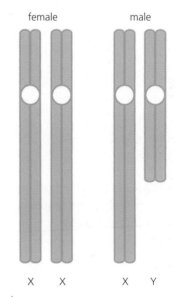

12.10 The sex chromosomes.

12.19 Sex is inherited.

You can work out sex inheritance in just the same way as for any other characteristic, but using the letter symbols to describe whole chromosomes, rather than individual alleles.

parents' phenotypes	female	male
parents' genotypes	XX	XY
gametes	X	X or Y

offspring genotypes

	X
X	XX
Y	XY

offspring phenotypes	XX = female	50%
	XY = male	50%

Half the children will probably be male, and half female.

Questions

1. If a normal human cell has 46 chromosomes, how many chromosomes are there in a human sperm cell?

2. Using the symbols W for normal wings, and w for vestigial wings, write down the following:

   (a) the genotype of a fly which is heterozygous for this characteristic

   (b) the possible genotypes of its gametes.

3. Using the layout shown in Section 12.14, work out what kind of offspring would be produced if the heterozygous fly in question 2 mated with one which was homozygous for normal wings.

4. In humans, the allele for red hair, b, is recessive to the allele for brown hair, B. A man and his wife both have brown hair. They have five children, three of whom have red hair, while two have brown hair. Explain how this may happen.

5. In Dalmatian dogs, the allele for black spots is dominant to the allele for liver spots. If a breeder has a black-spotted dog, how can she find out whether it is homozygous or heterozygous for this characteristic?

Variation and selection

12.20 Variation can be caused by genes.

We have seen how different alleles of a gene can produce variation in the phenotype of a person. For example, the alleles of the I gene that are present in your cells determine your blood group. The alleles that you have for the gene that codes for the membrane protein in your lungs determine whether you will have cystic fibrosis or not.

The examples of genetic crosses that we have worked through in this chapter have showing you how the offspring from two parents can be different from each other, and from their parents. This, of course, is only true for sexual reproduction. In asexual reproduction, where there is only one parent and no gametes are involved, the offspring are all genetically identical to each other and to their parents.

However, there is another way in which genes can cause variation. This can happen if there is a sudden change in the genes in a cell. Such a change, which is quite unpredictable, is called a **mutation**. A mutation can affect just one gene, or it might affect a whole chromosome.

One example of a mutation that affects a whole chromosome is the one involved in **Down's syndrome**. The mutation happens when an egg is being formed in a woman's ovaries. Instead of each egg cell getting one complete set of chromosomes (that is 23 altogether), one of them gets an extra chromosome 21. This egg therefore has 24 chromosomes altogether.

Practical 12.1 'Breeding' beads.

In this experiment, you will use two containers of beads. Each container represents a parent. The beads represent the gametes they make. The colour of a bead represents the genotype of the gamete. For example, a red bead might represent a gamete with genotype A, for dark hair. A yellow bead might represent a gamete with the genotype a, for fair hair.

1. Put 100 red beads into the first container. These represent the gametes of a person who is homozygous for dark hair, AA.

2. Put 50 red beads and 50 yellow beads into the second container. These represent the gametes of a heterozygous person with the genotype Aa.

3. Close your eyes, and pick out one bead from the first container, and one from the second. Write down the genotype of the 'offspring' they produce. Put the two beads back.

4. Repeat step 3 one hundred times.

5. Now try a different cross, for example Aa crossed with Aa.

Questions

1. In the first cross, what kind of offspring were produced, and in what ratios?

2. Is this what you would have expected? Explain your answer.

3. Why must you close your eyes when choosing the beads?

4. Why must you put the beads back into the containers after they have 'mated'?

When it is fertilised by a sperm with the correct number of 23 chromosomes, the zygote formed has 47 chromosomes. The child that develops from this zygote has an extra chromosome 21 in each of his or her cells. This affects the child's mental and physical development. Children with Down's syndrome usually have very happy and affectionate personalities, but they often have some degree of mental retardation, and are more likely to suffer from heart disease or infections.

An example of a mutation that affects just one gene is the one involved in **sickle cell anaemia**. The gene that mutates is one that codes for the production of haemoglobin, the red pigment found inside red blood cells that carries oxygen around the body. The change in the gene is very small, affecting just a tiny length of the DNA molecule. But this is enough to change the structure and function of the haemoglobin that it codes for. The haemoglobin that the cell makes is called **sickle cell haemoglobin**, and it is not very good at transporting oxygen. Even worse, when the oxygen concentration in the blood is low (for example, if the person has been doing vigorous exercise) the haemoglobin molecules inside the red blood cells form fibres that get tangled up with one another, and pull the cells into a sickle shape. These can easily get stuck inside blood capillaries, which is very painful and can be life-threatening.

Why do mutations happen? Some of them happen for no particular reason – something simply goes wrong inside a cell. This is most likely to happen when a cell is dividing, or getting ready to divide. The chance of a woman producing an egg with an extra chromosome 21 increases as she gets older. The chance of any kind of mutation happening is greatly increased if ionising radiation strikes the cell, because this can damage the DNA molecules in the chromosomes. Ultra-violet light can do this. There are also a number of different chemicals that can cause mutations, such as dioxins and mustard gas.

12.21 Variation can be caused by the environment.

Another important reason for variation is the difference between the **environments** of individuals. Scots pine trees possess genes which enable them to grow to a height of about 35 m. But if a Scots pine tree is grown in a very small pot, and has its roots regularly pruned, it will be permanently stunted (Fig 12.11). The tree's genotype gives it the potential to grow tall, but it will not realise this potential unless its roots are given plenty of space and it is allowed to grow freely.

Characteristics caused by an organism's environment are sometimes called acquired characteristics. They are not caused by genes, and so they cannot be handed on to the next generation.

A bonsai pine tree is dwarfed by being grown in a very small pot, and continually pruned.

Variation caused by the environment is not inherited.

A cutting from a bonsai pine would grow into a full size tree, if given sufficient space.

A dwarf pony, such as a Shetland pony, is small because of its genes. The offspring of Shetland ponies are small like their parents, no matter how well they are fed and cared for.

12.11 The inheritance of variation.

12.22 Variation can be continuous or discontinuous.

Some of the more obvious differences between people are in height or hair colour. We also vary in intelligence, blood groups, whether we can roll our tongues or not, and in many other ways.

There are two basic kinds of variation. One kind is **discontinuous variation**. Blood group is an example of discontinuous variation. Everyone fits into one of four definite categories – they have one of the four possible blood groups. There is no 'in between' category.

The other kind is **continuous variation**. Height is an example of continuous variation. There are no definite heights that a person must be. People vary in height, between the very lowest and highest extremes.

You can try measuring and recording discontinuous and continuous variation in Practical 12.2. Your results for continuous variation will probably look similar to Fig 12.13. This is called a **normal distribution**. Most people come in the middle of the range, with fewer at the lower or upper ends.

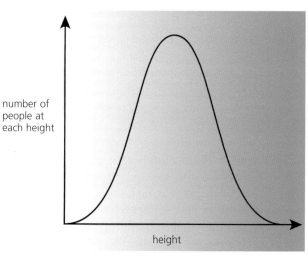

12.13 A normal distribution curve – a graph showing the numbers of people of different heights.

Questions

1 Explain the difference between genetic variation and environmental variation, giving examples of each.
2 Give one example of discontinuous variation.
3 Give one example of continuous variation.
4 What is most likely to be the cause of discontinuous variation – genes or the environment?
5 What is most likely to be the cause of continuous variation – genes or the environment?
6 What is a normal distribution?

12.12 Human height shows continuous variation. What characteristic here shows discontinuous variation?

Practical 12.2 Measuring variation

1 Make a survey of at least 30 people, to find out whether or not they can roll their tongue. Record your results.
2 Measure the length of the third finger of the left hand of 30 people. Take the measurement from the knuckle to the finger tip, not including the nail.
3 Divide the finger lengths into suitable categories, and record the numbers in each category, like this.

length / cm	number
8.0 – 8.4	2
8.5 – 8.9	4 and so on

Questions

1 Which of these characteristics is an example of continuous variation, and which shows discontinuous variation?
2 Your histogram may be a similar shape to the curve in Fig 12.13. This is called a **normal distribution**. The class which has the largest number of individuals in it is called the **modal class**. What is the modal class for the finger lengths of your samples?
3 The **mean** or average finger length is the total of all the finger lengths, divided by the number of people in your sample. What is the mean finger length of your sample?

Natural selection

12.23 Natural selection can cause gradual change.

Charles Darwin put forward a new and challenging theory in a book called *The Origin of Species*, which was published in 1859. Darwin's theory can be summarised as follows.

- **Variation**
  Most populations of organisms contain individuals which vary slightly from one another. Some slight variations may better adapt some organisms to their environment than others.

- **Over-production**
  Most organisms produce more young than will survive to adulthood.

- **Struggle for existence**
  Because populations do not generally increase rapidly in size there must therefore be considerable competition for survival between the organisms (Fig 12.15).

- **Survival of the fittest**
  Only the organisms which are really well adapted to their environment will survive.

- **Advantageous characteristics passed on to offspring**
  Only these well adapted organisms will be able to reproduce successfully, and will pass on their advantageous characteristics to their offspring.

- **Gradual change**
  In this way, over a period of time, the population will lose all the poorly adapted individuals. The population will gradually become better adapted to its environment.

This theory is called the **theory of natural selection**, because it suggests that the best adapted organisms are selected to pass on their characteristics to the next generation (Fig 12.16, overleaf).

Darwin proposed his theory before anyone understood how characteristics were inherited. Now that we know something about genetics, his theory can be stated slightly differently. We can say that natural selection results in the genes producing advantageous phenotypes being passed on to the next generation more frequently than the genes which produce less advantageous phenotypes.

12.14 A portrait of Charles Darwin at the age of 72.

12.15 When large numbers of organisms, such as these wildebeest of the East African plains, live together, there is competition for food, and a tendency for the weaker ones to be killed by predators. The organisms best adapted to their environment are most likely to survive.

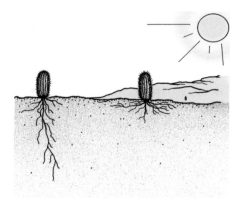
1 Genetic variation In a population of cacti, some have longer roots than others.

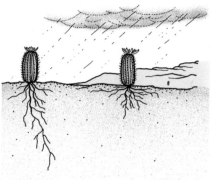
In the wet season they flower.

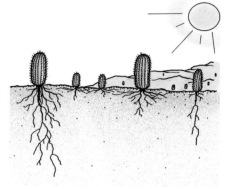
2 Overproduction The cacti produce large numbers of offspring.

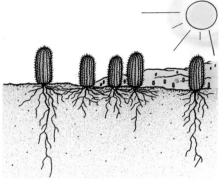
3 Struggle for existence During the dry season, there is competition for water.

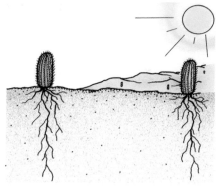
4 Survival of the fittest The cacti with the longest roots are able to obtain water, while the others die from dehydration.

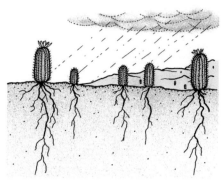
5 Advantageous characteristics passed on to offspring When conditions are suitable, the long-rooted cacti reproduce, producing long-rooted offspring.

12.16 An example of how natural selection might occur.

12.24 Melanic moths are selected near cities.

Darwin's theory of natural selection provides a good explanation for our observations of the many types of animals and plants. For example, it can help us to understand some changes that have taken place in a species of moth in Britain.

The peppered moth, *Biston betularia*, lives in most parts of Britain. It flies by night, and spends the daytime resting on tree trunks. It has speckled wings, which camouflage it very effectively on lichen-covered tree trunks (Fig 12.17).

People have collected moths for many years, and we know that up until 1849 all the moths in collections were speckled. But in 1849, a black or **melanic** form of the moth was caught near Manchester. By 1900, 98% of the moths near Manchester were black.

The distribution of the black and speckled forms in 1958 is shown in Fig 12.18.

How can we explain the sudden rise in numbers of the dark moths, and their distribution?

We know that the black colour of the moth is caused by a single dominant gene. The mutation from a normal to a black gene happens fairly often, so it is reasonable to assume that there have always been

12.17a Lichen-covered bark hides a speckled moth.

12.17b Dark moths are better camouflaged on lichen-free trees.

a few black moths around, as well as pale speckled ones. Up until the mid-nineteenth century, the pale moths had the advantage, as they were better camouflaged on lichen-covered tree trunks.

But in the middle of the nineteenth century, the Industrial Revolution took place. Some areas became polluted by smoke. As the prevailing winds in Britain blow from the west and south-west, the worst affected areas were to the east of industrial cities like Manchester and Birmingham. The polluted air prevented lichens from growing. Dark moths were better camouflaged than pale moths on trees with no lichens on them.

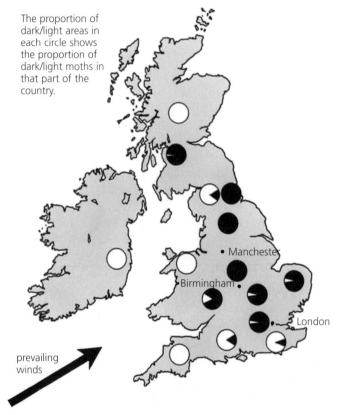

The proportion of dark/light areas in each circle shows the proportion of dark/light moths in that part of the country.

prevailing winds

12.18 The distribution of the pale and dark forms of the peppered moth, *Biston betularia*, in 1958. Since then, the number of dark moths has been decreasing, because there is now less air pollution.

So, in polluted areas, dark peppered moths were less likely to be eaten by birds than speckled peppered moths. The dark moths were more likely to survive long enough to breed, and their offspring inherited the genes for dark colour. Not many speckled moths had offspring, so quite soon there were not many speckled moths at all – they were mostly black ones.

Towards the end of the twentieth century, strict pollution controls had greatly reduced the amount of soot and other pollutants in the air. Tree bark was no longer covered with soot, and lichens could grow again. Now the speckled moths had the advantage again.

They were more likely to survive and breed, and pass on their genes to their offspring. Today, most peppered moths are once again of the speckled variety.

The factor which confers an advantage on one variety of peppered moths, and a disadvantage on other varieties, is predation by birds. This is called a **selection pressure**, because it 'selects' one variety for survival.

12.25 Antibiotic resistance in bacteria is selected.

Another example of natural selection can be seen in the way that bacteria may become resistant to antibiotics, such as penicillin. Penicillin works by stopping bacteria from forming cell walls. When a person infected with bacteria is treated with penicillin, the bacteria are unable to grow new cell walls, and they burst open.

However, the population of bacteria in the person's body may be several million. The chances of any one of them mutating to a form which is not affected by penicillin is quite low, but because there are so many bacteria, it could well happen. If it does, the mutant bacterium will have a tremendous advantage. It will be able to go on reproducing while all the others cannot. Soon, its descendants may form a huge population of penicillin-resistant bacteria.

This does, in fact, happen quite frequently. This is one reason why there are so many different antibiotics available – if some bacteria become resistant to one, they may be treated with another.

The more we use an antibiotic, the more we are exerting a selection pressure which favours the resistant forms. If antibiotics are used too often, we may end up with resistant strains of bacteria which are very difficult to control.

12.26 People heterozygous for sickle cell anaemia are selected for.

Sickle cell anaemia is an inherited disease. You can read more about it in Chapter 13. It is caused by an allele, H^S. People with the genotype H^AH^A or H^AH^S are normal, while people with the genotype H^SH^S have sickle cell anaemia. It is an unpleasant disease, and many people with it die before they can have children.

For some time, people were puzzled why sickle cell anaemia is common in some parts of the world. In some parts of Africa, for example, as many as 14% of the babies born have sickle cell anaemia. Why has natural selection not eliminated the H^S allele from the population?

If you look at the maps in Fig 12.19, you can see that the places where sickle cell anaemia is found match quite closely with the distribution of another disease – malaria. Malaria is another potentially fatal disease, which you can read more about in Chapter 13.

It was found that people with the genotype $H^A H^S$ were far less likely to suffer and die from malaria than people with the genotype $H^A H^A$. So there are different selection pressures acting on people with the three different genotypes. People with the genotype $H^A H^A$ are at risk of dying from malaria if they live in certain parts of the world. People with the genotype $H^S H^S$ are at risk of dying from sickle cell anaemia. People with the genotype $H^A H^S$ have a strong selective advantage in areas where malaria is present, because they are more likely to live and reproduce than people with the other two genotypes.

Therefore, in each generation, the people most likely to reproduce are heterozygous people. Some of their children will also be heterozygous, but some will be homozygous dominant, and some homozygous recessive. This will continue generation after generation – until someone finds a really good cure or prevention for malaria, or a really successful treatment for sickle cell anaemia.

12.27 Natural selection can result in evolution.

Most biologists believe that natural selection has had an important part to play in **evolution**.

Evolution can be defined as a change in the characteristics of a species over time. If a mutation produced new and advantageous features in an individual organism, then natural selection would mean that this organism would be more likely to survive than all the others that did not have this feature. Its offspring could inherit this feature, and in turn pass it on to their offspring. If this happened time and time again, over very long periods of time, then we can imagine how organisms looking very different from their ancestors could arise.

Unfortunately, we cannot look back in time to find out if this really is how evolution happened. But we can watch natural selection happening now – for example, the changes in the peppered moths described in Section 12.24. Logically, it makes sense to suggest that this process has happened in the past, too. Natural selection may not have been the only process by which evolution has taken place, but it has almost certainly had a very important role to play.

12.28 Natural selection does not always cause change.

Natural selection does not always produce change. Natural selection ensures that the organisms which are best adapted to their environment will survive. Change will only occur if the environment changes, or if a new mutation appears which adapts the organism better to the existing environment.

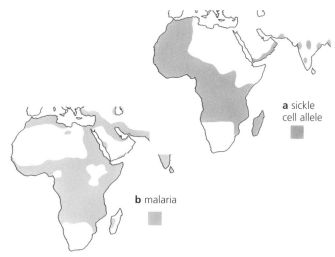

12.19 The distribution of **a** the sickle cell allele and **b** malaria.

For example, in the south-west of Britain the environment of the peppered moth has never changed very much. The air has not become polluted, so lichens have continued to grow on trees. The best camouflaged moths have always been the pale ones. So selection has always favoured the pale moths in this part of Britain. Any mutant dark moths which have appeared have always been at a disadvantage, and unlikely to survive.

Most of the time, natural selection tends to keep populations very much the same from generation to generation. It is sometimes called **stabilising selection**. If an organism is well adapted to its environment, and if that environment stays the same, then the organism will not evolve. The strange lobe-finned fish, the coelacanth, for example, has remained virtually unchanged for 350 million years. Coelacanths live deep in the Indian Ocean, which is a very stable environment (Fig 12.20)

12.20 Coelacanths, which live deep in the Indian Ocean, have existed almost unchanged for 350 million years. Humans have existed for only about 4 million years.

12.29 Humans have used artificial selection for thousands of years.

Humans can also bring about changes in living organisms, by selecting certain varieties for breeding. Figs 12.21 and 12.22 show examples of the results of this kind of selection. From the varied individuals amongst a herd of cattle, the breeder chooses the ones with the characteristics he wants to appear in the next generation. He then allows these individuals, and not the others, to breed. Over many generations, these characteristics will become the most common ones in the population.

This process is called **artificial selection**. It has been going on for thousands of years, ever since humans first began to cultivate plants and to domesticate animals. It works in just the same way as natural selection. Individuals with 'advantageous' characteristics breed, while those with 'disadvantageous' ones do not.

However, what humans think are desirable characteristics would often not be at all advantageous to the plant or animal if it was living in the wild. Modern varieties of cattle, for example, selected over hundreds of years for high milk yield or fast meat production, would stand little chance of surviving for long in the wild.

Some farmers are now beginning to think differently about the characteristics they want in their animals and plants. Instead of enormous yields as their first priority, they are now looking for varieties which can grow well with less fertiliser or pesticide in the case of food plants, and with less expensive housing and feeding in the case of animals. Luckily, many of the older breeds, which had these characteristics have been conserved, and can now be used to breed new varieties with 'easy-care' characteristics.

Questions

1 Using the six points listed in Section 12.23, explain why the proportion of dark peppered moths near Manchester increased at the end of the nineteenth century.

2 Imagine you are a farmer with a herd of dairy cattle. You want to build up a herd with a very high production of milk. You have access to sperm samples from bulls, for each of which there are records of the milk production of his offspring. What will you do?

3 Wheat is attacked by many different pests, including a fungus called yellow rust.
(a) Describe how you could use artificial selection to produce a new variety of wheat which is naturally resistant to yellow rust.
(b) How could the growing of resistant varieties reduce pollution?
(c) When resistant varieties of wheat are produced, it is found that after a few years they are infected by yellow rust again. Explain how this might happen.

12.22a Longhorn Ankole cattle, like these photographed in Africa, are a very old breed. They are thought to be quite similar to original wild cattle.

12.22b Friesian cattle, shown in the background, have been bred for high milk yield. Jersey cattle, shown in the foreground, have been bred for high fat content of milk.

a Crab apple (wild)

b James Grieve apple (cultivated)

12.21 Wild and cultivated apples.

13 Health, disease and medicine

13.1 There are many causes of disease.

People often think of a disease as being something caused by bacteria or viruses, like flu (influenza) or AIDS. While many diseases *are* caused by microorganisms, there are many other causes of disease. The most important ones are listed below:

1. Genetic diseases, caused by **genes**. These include cystic fibrosis, sickle cell anaemia, Down's syndrome and some kinds of diabetes.
2. Infectious diseases caused by other living organisms, which are called **pathogens**. These include influenza, AIDS, malaria, cholera, syphilis, gonorrhoea and many others.
3. Autoimmune diseases, in which a person's **immune system** begins to attack the body's own cells. These include some kinds of diabetes, some kinds of arthritis, and muscular dystrophy.
4. Self-inflicted diseases, where a person damages their health by a poor **life-style**. These include alcoholism, drug addiction and most cases of lung cancer.
5. Deficiency diseases, caused by a lack of a particular nutrient in a **diet**. These include scurvy and anaemia, caused by a lack of vitamin C and iron respectively.
6. Degenerative diseases, caused by natural **ageing** processes. These include most heart disease, and some kinds of arthritis.

With so many kinds of disease, you might wonder how anyone ever manages to stay well! However, many of these diseases are avoidable or treatable – or both. In this chapter, we will look briefly at one genetic disease, and then concentrate on diseases caused by pathogens. We will also consider how your life-style affects your chances of suffering from serious diseases.

Questions
List some foods that could prevent
(a) scurvy — lack of Vitamin C
(b) anaemia. lack of Iron

Genetic diseases

13.2 Sickle cell anaemia is caused by a recessive allele.

Sickle cell anaemia is caused by an allele of the gene which codes for the production of haemoglobin. The sickle cell allele is recessive, so only people who are homozygous for this allele have the disease.

The sickle cell allele causes the production of abnormal haemoglobin, which does not carry oxygen very well. When the concentration of oxygen in the blood becomes low – for example, if the person is doing exercise – the sickle cell haemoglobin forms a precipitate inside the red blood cells. It pulls the red cells into a curved, sickle shape instead of their normal round shape (Fig 13.1). The cells get stuck in capillaries, which is very painful and stops blood flowing through them. Cells get starved of oxygen, which is why the disease is called 'anaemia'.

If you look back at Section 12.26, you will see that sickle cell anaemia is most common in parts of Africa where malaria is present. Sickle cell anaemia is now also found in other parts of the world, such as Britain and the USA, because people whose ancestors came from malarial regions now live there.

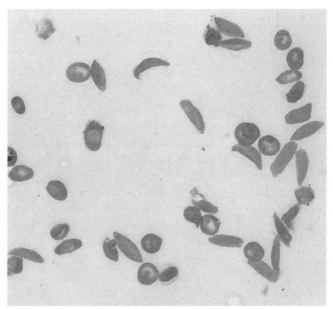

13.1 This sample of blood contains some normal, round red blood cells, and some which have 'sickled'. You can imagine how easily the sickle-shaped ones could get stuck in a narrow capillary.

Infectious diseases

13.3 A pathogen is an organism which causes disease.

Many diseases are caused by other living organisms, which get into our bodies and breed there. These organisms are called **pathogens**. Pathogens may damage our cells directly, by living in them and feeding on them. They may produce waste products called **toxins**, which spread around the body in the blood and cause symptoms such as high temperature and rashes. Some toxins – such as the one produced by the bacterium *Clostridium botulinum* – are amongst the most dangerous poisons in the world.

Pathogens may belong to one of four different groups of organisms – viruses, bacteria, protoctists and fungi. Table 13.1 shows some diseases caused by organisms from each of these groups.

13.4 Pathogens get into the body in different ways.

There are several ways in which pathogens can get into your body.

Through the skin

The skin usually makes a very good protective barrier around your body, stopping most pathogens from gaining entry. However, there are a few bacteria and viruses which can get through undamaged skin. One of these is the virus which causes warts. The bacterium *Staphylococcus* can get in through a cut, turning it septic.

Through the respiratory passages

Cold and influenza viruses are carried in the air in tiny droplets of moisture. Every time someone with these illnesses speaks, coughs or sneezes, millions of viruses are propelled into the air. If you breathe in the droplets, you may become infected.

In food or water

Bacteria such as *Salmonella* can enter your alimentary canal with food you eat. If you eat a large amount of them, you will get food poisoning. Many pathogens, including the virus which causes polio and the bacterium which causes cholera, are transmitted in water. If you drink untreated water, you run the risk of catching these diseases.

By vectors

In Biology, a vector is an organism which carries a pathogen from one host to another. One example of a vector is the mosquito which transmits malaria. The mosquito injects the malarial pathogen (which is a protoctist) into a person's blood when it bites.

13.5 Influenza is caused by a virus.

Influenza is a very infectious disease – this means that it can easily be passed from one person to another. The virus which causes it is usually breathed in in tiny droplets of moisture.

Viruses are very strange organisms. Indeed, some people think they should not be classified as living things at all! They are not made of cells, but just some protein molecules around some DNA or RNA (Fig 13.2). They cannot do any of the things which living things are supposed to be able to do, such as respire or grow. They cannot do anything at all until they get inside another living cell – when they begin to reproduce.

Flu viruses get into the cells lining your respiratory passages. They take over your cells, and instruct them to make more flu viruses. So many new viruses are made that the cell is destroyed. The new flu viruses burst out of the cell, ready to infect more of your cells, or to be breathed out in droplets of moisture, ready to infect someone else's cells.

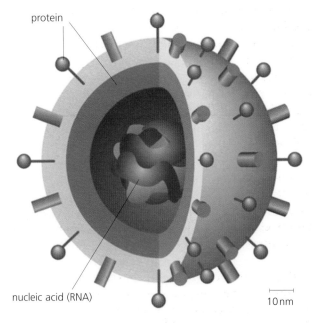

13.2 A flu virus.

Table 13.1 Types of pathogens.

Group to which pathogens belong	Examples of diseases caused
Viruses	Influenza, common cold, poliomyelitis, AIDS
Bacteria	Cholera, syphilis, gonorrhoea, tuberculosis, tetanus
Protoctists	Malaria, amoebic dysentery
Fungi	Athlete's foot, ringworm

Not surprisingly, you feel rather ill while this is happening. For the first three days after **infection** – the time when you first breathed the viruses in – nothing very much happens. This is called the **incubation period**. Then your temperature starts to go up, and you will probably get aching muscles, a headache, a sore throat and perhaps a cough.

If you are relatively fit and healthy, your immune system will quite quickly destroy all the infected cells and the viruses, so that you will recover within about one or two weeks. The way the immune system does this is described in Sections 13.9–13.11. However, if someone has a weak immune system, then other pathogens may take advantage of the viral infection to get a toehold in the respiratory system. They can cause **secondary infections**, producing bronchitis and pneumonia. These can be fatal. Major outbreaks of influenza often cause death in some old people.

Another disease that is caused by a virus is acquired immune deficiency syndrome, AIDS. This is described in Section 13.14.

13.6 Gonorrhoea and syphilis are sexually transmitted diseases.

Both gonorrhoea and syphilis are caused by bacteria that can be passed from one person to another during sexual intercourse.

The bacterium that causes gonorrhoea is a small, round cell. It can only survive in moist places, such as the tissues lining the tubes in the reproductive systems of a man or a woman. If gonorrhoea bacteria are living in a woman's vagina or a man's urethra, the infection can be passed from one to the other during sexual intercourse.

The first symptoms usually occur between two and seven days after infection. In a man, the bacteria reproduce inside the urethra, and this produces an unpleasant discharge and pain when urinating. In a woman, the bacteria reproduce mostly in the cervix, although they can also do so in the vagina. As in men, this produces a discharge, but many women do not notice this and they do not suffer pain as men do. Consequently, while most men with gonorrhoea know that they have it, many women are quite unaware that they have the infection.

The bacterium that causes syphilis is a spiral-shaped cell. It is transmitted from person to person in just the same way as gonorrhoea. The first symptoms usually occur about one or two weeks after infection, in the form of a painless ulcer at the place where the syphilis bacteria were introduced into the body. This ulcer disappears after about two months, and many people take this to mean that there is nothing wrong with them. However, the bacteria now spread all over the body, and after a while the person may suffer from a sore throat, headache and fever, followed by a skin rash. Once again, these symptoms all disappear, so once again the person may think that they are well. Indeed, some of them may be, and do not have any other problems from their infection with syphilis. In many people, however, the bacteria continue to live in the body for many years. After some years, the bacteria may begin to destroy almost any organ in the body, such as the heart or the brain.

Both gonorrhoea and syphilis can be treated with antibiotics (Section 13.17) such as penicillin, and this is almost always successful. However, it is much better to make sure that you do not get either of these diseases in the first place! This can be achieved by:

- not being sexually active
- having only one sexual partner – if neither has the disease, then they cannot pass it on to each other
- ensuring that the man uses a condom, as the bacteria cannot pass through this from one person to the other
- tracing, warning and treating all possible sexual contacts of a person who is diagnosed with gonorrhoea or syphilis, to make sure that it does not spread any further.

13.7 Malaria is transmitted by a vector.

The pathogen which causes malaria is a protoctist called *Plasmodium* (Figs 13.3 and 13.4). Malaria is one of the world's biggest killers. Up to 500 million people have malaria every year. The distribution of malaria in Africa is shown in Fig 12.19.

You get malaria by being bitten by a female mosquito belonging to the genus *Anopheles* if she is carrying *Plasmodium* in her saliva. The mosquito has a long proboscis, which she pushes through the skin into a blood vessel. She sucks up blood through the proboscis. To stop the blood clotting, she injects some saliva. If her saliva contains *Plasmodium* then she injects *Plasmodium* as well. The mosquito is a **vector** for the disease. This explains the distribution of malaria in the world – you find malaria where you find the mosquitoes which transmit it.

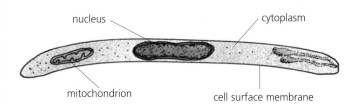

13.3 *Plasmodium*, the protoctist that causes malaria.

The *Plasmodium* organisms get into your liver cells. They breed and, after a few days, leave the liver and enter some of your red blood cells. This is when you first feel ill, feeling a bit as though you have flu.

The organisms breed even faster when they are in the red blood cells. When a red cell becomes full of young organisms, it bursts, releasing them into your blood so that they can infect other red blood cells. This happens at regular intervals – say every two days. The exact interval depends on the species of *Plasmodium* with which you are infected.

Every time the red cells burst, your temperature goes up and you feel ill. This stage of the disease can be very dangerous, and many people die. Body temperature may go up to 40 °C. The burst blood cells may block blood vessels, stopping oxygen supplies getting to many parts of the body.

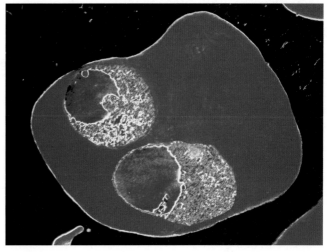

13.5 Electron micrograph of a red cell from a person suffering from malaria. The red cell contains two `Plasmodium` cells, which have been coloured blue.

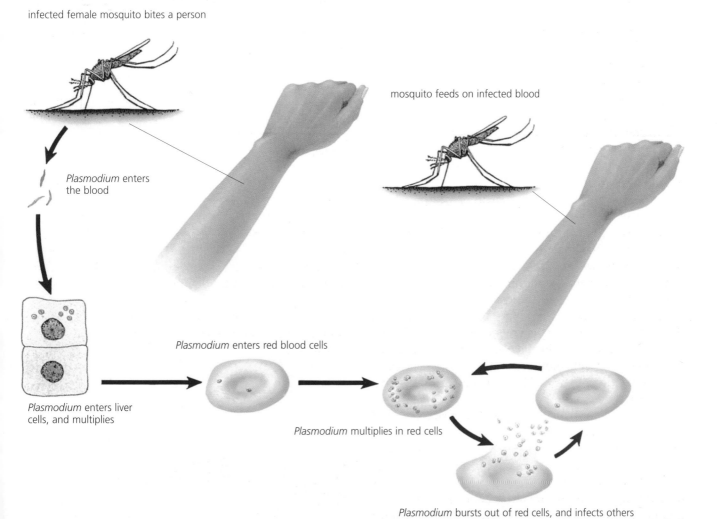

13.4 How malaria is transmitted.

13.8 Malaria is difficult to control.

Malaria has proved very difficult to control. There are several different lines of attack which have been tried.

The life cycle of the mosquito which transmits malaria is shown in Fig 13.6. If mosquitoes can be killed at some stage of this life cycle, then the number of mosquitoes available to transmit malaria should be reduced.

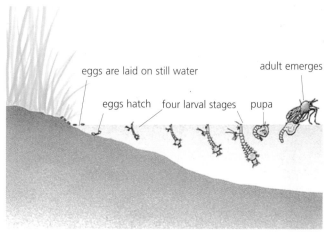

13.6 The life cycle of *Anopheles*, a mosquito which transmits malaria.

Insecticides

Adult mosquitoes are active at night, and rest in places such as houses during the day. **Insecticides**, including DDT, have been used to spray their resting places. This has been very effective in some areas, but in many places the mosquitoes have now evolved to become resistant to insecticides.

Dealing with ponds

You can see from Fig 13.6 that mosquitoes lay their eggs in stagnant water, and the larvae live in the water before developing into adults. They have to come to the surface to get oxygen from the air. In some places, swamps and open water have been drained, to reduce the amount of water available for mosquitoes to breed. People take care not to leave containers of water, or even puddles, near their houses. Oil can be sprayed onto open water to make a film on the surface, preventing the larvae from coming to the surface to get air. All of these methods have had some success, but they are often too expensive to carry out often enough to destroy all the larvae, and they can also harm other organisms.

Biological control

Fish have been put into ponds where mosquitoes breed, to eat the larvae. Bacteria which infect the larvae have been put into the water. These methods do less harm to other organisms, but they do not kill all of the larvae.

Avoiding mosquitoes

Malaria can only be transmitted if female mosquitoes bite an infected person, and then an uninfected one. People should sleep under a **mosquito net** (Fig 13.7), so that they cannot be bitten. Insect-repellent creams put onto the skin can also help, but they do not completely prevent you being bitten.

As the mosquitoes that transmit malaria are most active at night, it helps to wear clothes that fully cover your arms and legs during the evenings.

Drugs

Another approach is to use **drugs** to kill the *Plasmodium* inside an infected person's body. There are several different drugs available. There are also drugs which you can take to prevent the *Plasmodium* ever infecting you, even if you are bitten by a mosquito carrying *Plasmodium*. People visiting a country where malaria is present should always take these drugs. However, *Plasmodium* appears to be able quickly to become resistant to them, and new drugs are always having to be developed.

Until recently, there was no **vaccine** available for malaria. However, in 1994, a new vaccine proved successful in trials in South America and Africa.

13.7 Sleeping under a mosquito net is a good protection against malaria. The mosquitoes which are vectors for the disease are most active after dark.

Body defences against infectious disease

13.9 The body has many defences against infection.

The human body has many different ways of stopping pathogens from getting into it. Table 13.2 lists some of these. However, you can see from this Table that we do not have any natural defences against the entry of the pathogens that cause malaria, or those that cause sexually transmitted diseases such as gonorrhoea, syphilis or AIDS.

If pathogens do get past these defences and into the body, then our **white blood cells** come into action and attempt to destroy them.

There are many different kinds of white blood cells. They all have the function of destroying pathogens in your body, but they do it in different ways.

Phagocytes are cells which can move around the body, engulfing and destroying pathogens (Fig 13.8). They also destroy any of your own cells which are damaged or worn out. Phagocytes often have lobed nuclei. If you damage your skin, perhaps with a cut or graze, phagocytes will collect at the site of the damage, to 'mop up' any microorganisms which might possibly get in.

Lymphocytes have a quite different method of attacking pathogens. They produce chemicals called **antibodies**, which are carried in the blood and tissue fluid to almost every part of the body.

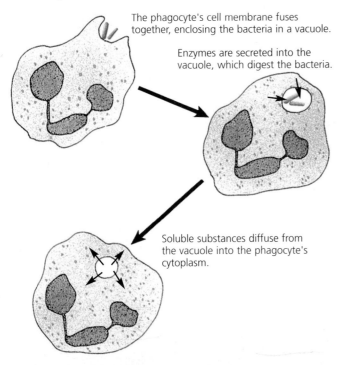

13.8 A white cell destroying bacteria by phagocytosis.

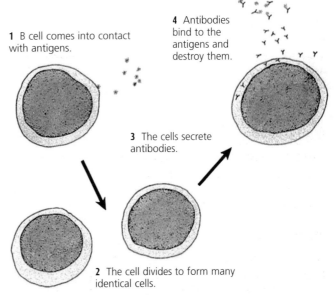

13.9 How B lymphocytes respond to antigens.

Table 13.2 How the body prevents infection.

Method of entry	Example	Natural defences
Through skin	*Staphylococcus* bacterium	• Epidermis is a barrier between pathogens and body • When skin is damaged, blood clots seal wound and prevent entry of pathogens (Section 7.16) • Tears contain lysozyme, which helps to prevent eye infections
Into respiratory system	Influenza virus	• Cilia and mucus in respiratory passages trap dust particles, which may carry pathogens, and sweep them upwards (Section 6.6) • Phagocytes patrol the surfaces of the alveoli, and destroy any pathogens they find
In food or water, into alimentary canal	*Salmonella*	• Distaste for food which looks or smells bad • Hydrochloric acid in stomach kills many bacteria (Section 4.36)
Injection into body by a vector	*Plasmodium*	None
By sexual intercourse	HIV	None

13.10 Antibodies are specific.

In your body, you have thousands of different kinds of lymphocytes. Each kind is able to produce a different sort of antibody.

An antibody is a protein molecule with a particular shape. Rather like an enzyme molecule, this shape is just right to fit into another molecule. To destroy a particular pathogen, antibody molecules must be made which are just the right shape to fit into molecules on the outside of the pathogen. These pathogen molecules are called **antigens**.

When antibody molecules lock onto the pathogen, they kill the pathogen. There are several ways in which they do this. One way is simply to alert phagocytes to the presence of the pathogen, so that the phagocytes will come and destroy them. Or they may start off a series of reactions in the blood which produce enzymes to digest the pathogens.

13.11 Lymphocytes multiply when 'their' pathogen is present.

Most of the time, most of your lymphocytes do not produce antibodies. It would be a waste of energy and materials if they did. Instead, each lymphocyte waits for a signal that a pathogen which can be destroyed by its particular antibody is in your body.

If a pathogen enters the body, it is likely to meet a large number of lymphocytes. One of these may recognise the pathogen as being something that its antibody can destroy. This lymphocyte will start to divide rapidly by mitosis, making a clone of lymphocytes just like itself. These lymphocytes then secrete their antibody, destroying the pathogen (Fig 13.9).

This takes time. It may take a while for the 'right' lymphocyte to recognise the pathogen, and then a few days for it to produce a big enough clone to make enough antibody to kill the pathogen. In the meantime, the pathogen breeds, making you ill. Eventually, however, the lymphocytes get the upper hand, and you get better (Fig 13.10).

The type of lymphocytes that release antibodies are called **B lymphocytes**. There is another kind of lymphocyte, called **T lymphocytes**, that behave slightly differently. Just like B lymphocytes, they divide and produce a clone when they meet 'their' particular pathogen. But, instead of releasing antibodies, T lymphocytes attach themselves directly to the pathogens, or to body cells that are infected by pathogens. The T lymphocytes then destroy them. This means that T cells are especially good at destroying viruses, because viruses live and breed inside our body cells.

Lymphocytes are a very important part of your **immune system**. The way in which they respond to pathogens, by producing antibodies or directly attacking them, is called the **immune response**.

13.12 Memory cells make you immune.

The first time a pathogen enters your body, it takes a little while for the immune response to swing fully into action. This gives the pathogen a chance to breed, so you get ill.

However, when the lymphocytes multiply to form clones, not all of the cells in the clone take on the job of making antibodies. Some of them remain inactive. They are called **memory cells**. Both B and T cells produce memory cells. Memory cells stay in your blood for a very long time after an infection.

If the same pathogen gets into your body again, it is likely to be recognised by a memory cell almost straight away. The immune response to the pathogen is immediate, killing it before it has any chance to breed. You are now immune to that disease.

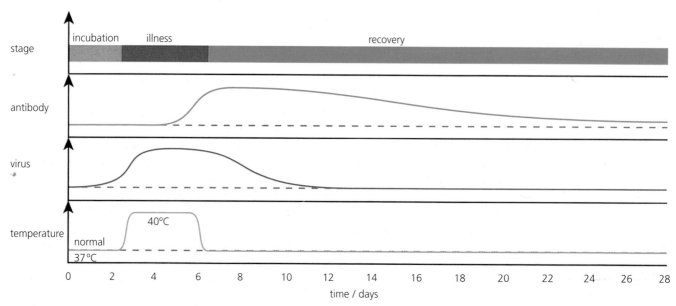

13.10 How the number of viruses and antibodies in your body and your body temperature, change during a bout of flu.

13.13 Cold viruses keep changing.

For many diseases, immunity lasts all your life. If you have mumps or measles once, you will probably never have it again. However, this is not true for one of the most common diseases – the common cold.

Colds are caused by a virus. The special feature of cold viruses which helps them to win the battle against our immune response is that they keep changing the proteins in their outer coat. If you get a cold, the lymphocytes which make an antibody against the particular antigens of the cold virus which is infecting you form memory cells. But then along comes another cold virus with a slightly different protein coat. The memory cells do not recognise it. You get a cold all over again, and produce another set of memory cells. Then along comes another virus. You can't win!

13.14 AIDS is caused by a virus which attacks lymphocytes.

The disease **AIDS**, or acquired immune deficiency syndrome, is caused by **HIV**. HIV stands for **human immunodeficiency virus**. Fig 13.11 shows this virus.

HIV infects lymphocytes, in particular T cells. Over a long period of time, HIV slowly destroys T cells. Several years after infection with the virus, the level of certain kinds of T cells is so low that they are unable to fight against other pathogens effectively. Because HIV attacks the very cells which would normally kill viruses – the T cells – it is very difficult for someone's own immune system to protect them against HIV.

About ten years after initial infection with HIV, a person is likely to develop symptoms of AIDS. They become very vulnerable to other infections, such as pneumonia. They may develop cancer, because one function of the immune system is to destroy body cells which may be beginning to produce cancers. Brain cells are also quite often damaged by HIV. A person with AIDS usually dies of a collection of several illnesses.

There is still no cure for AIDS. Researchers are always trying to develop new drugs, which will kill the virus without damaging the person's own cells. As yet, no vaccine has been produced either, despite large amounts of money being spent on research.

13.15 HIV is transmitted in body fluids.

The virus which causes AIDS cannot live outside the human body. In fact, it is an especially fragile virus – much less tough than the cold virus, for example. You can only become infected with HIV through direct contact of your body fluids with those of someone with the virus. This can be in one of the following ways.

Through sexual intercourse

HIV can live in the fluid inside the vagina, rectum and urethra. During sexual intercourse, fluids from one partner will come into contact with fluids of the other. It is very easy for the virus to be passed on in this way.

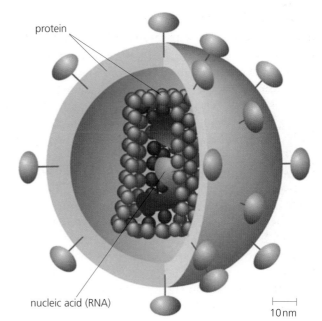

13.11 The human immunodeficiency virus, HIV.

The more partners a person has, the higher the chance of them becoming infected with HIV. In some parts of the world, where it is common practice for men to have many different sexual partners, extremely high percentages of people have developed AIDS. This is so in some parts of Africa and Asia, and also amongst some homosexual communities in parts of Europe and the USA.

The best way of avoiding AIDS is never to have more than one sexual partner. If *everyone* did that, then AIDS would immediately stop spreading. Using condoms is a good way of lowering the chances of the viruses passing from one person to another during sexual intercourse – though it does not rule it out.

Through blood contact

Many cases of AIDS have been caused by HIV transferred from one person's blood to another. In the 1970s and 1980s, when AIDS first appeared, and before anyone knew what was causing it, blood containing HIV was used in transfusions. People being given the transfusions were infected with HIV, and later developed AIDS. Now all blood used in transfusions in most countries is screened for HIV before it is used.

Blood can also be transferred from one person to another if they share hypodermic needles. This most commonly happens in people who inject drugs, such as heroin. Many drug users have died from AIDS.

People who have to deal with accidents, such as police and ambulance attendants, must always be on the guard against AIDS if there is blood around. They often wear protective clothing, just in case a bleeding accident victim is infected with HIV.

However, in general, there is no danger of anyone becoming infected with HIV from contact with someone with AIDS. You can quite safely talk to them, shake hands with them, drink from cups which they have used and so on. In fact, there is far more danger to the person who *has* AIDS from such contacts, because they are so vulnerable to any bacterium or virus which they might catch from you.

13.16 The immune system can reject transplants.

People suffering from serious diseases affecting a particular organ may be given a **transplant**. Many different organs can now be transplanted. Some of the most common transplants are kidney transplants and bone marrow transplants. Heart transplants are also given quite frequently.

The person receiving the transplant is the **recipient**, and the person from whose body the organ was taken is the **donor**. Often, the donor is someone who died in an accident. Many people carry donor cards with them all the time, stating that they are happy for their organs to be used in a transplant operation. Organs for transplants must be removed quickly from a body and kept cold, so that they do not deteriorate. Sometimes, however, the donor may be alive. A person may donate a kidney to a brother or sister who needs one urgently. You can manage perfectly well with just one kidney.

Surgeons now have very few problems with transplant operations – they can almost always make an excellent job of removing the old organ and replacing it with a better one. The big problem comes afterwards. The recipient's immune system recognises the donor organ as being foreign, and attacks it. This is called **rejection**.

The recipient is given drugs called **immunosuppressants** which stop the white blood cells working efficiently, to decrease the chances of rejection. The trouble with immunosuppressants is that they stop the immune system from doing its normal job, and so the person is more likely to suffer from all sorts of infectious diseases. The drugs have to be taken for the rest of the recipient's life.

The chances of rejection are reduced if the donor is a close relative of the recipient. This, of course, can only be considered when the organ for donation is one which can be given by a living person, such as a kidney or some bone marrow (Fig 12.12). Closely-related people are more likely to have antigens on their cells which are similar to each other, so the recipient's immune system is less likely to react to the donated organ as if it were 'foreign'. If there is not a relative who can donate an organ, then a search may be made world-wide, looking for a potential donor with similar antigens to the recipient.

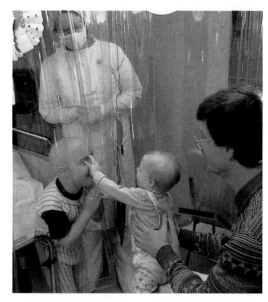

13.12 This little boy suffers a disease in which his bone marrow does not make enough red cells. He was given a transplant of blood, containing bone marrow cells, taken from the umbilical cord of his baby sister.

13.17 Antibiotics can kill bacteria in the body.

Sometimes, a person's body needs help in its fight against a bacterial infection. Until 1944, there was little help which could be given. People died from diseases which we now think quite harmless, such as infected cuts.

Then a discovery was made which has had a tremendous effect on our ability to treat diseases. **Antibiotics** were discovered.

Antibiotics are substances which kill bacteria, but do not harm living cells. Most of them are made by fungi. It is thought that the fungi make antibiotics to kill bacteria living near them – bacteria and fungi are both saprophytes, so they might compete for food. We use the chemical warfare system of the fungus to wage our own war against bacteria.

The first antibiotic to be discovered was **penicillin**. It is made by the fungus *Penicillium*, which you might sometimes see growing on decaying fruit. Penicillin kills bacteria by stopping them making their cell walls. Since the introduction of penicillin, many more antibiotics have been found.

We have to go on trying to find more and more antibiotics, because bacteria evolve to become resistant to them, as described in Section 12.25. The more we use antibiotics, the more selection pressure we put on bacteria to evolve resistance. People did not realise this when antibiotics were first discovered, and used them for all sorts of diseases where they did not help at all, such as diseases caused by viruses. Now doctors are much more careful about the amount of antibiotics which they prescribe. We should only use antibiotics when they are really needed – then there is more chance that they will work when we need them to.

Life-style and disease

13.18 Life-style affects health.

In many countries today, a major cause of death is **coronary heart disease**. **Cancer** and **strokes** are also important. To some extent, these diseases are all caused by natural ageing processes. However, they are also affected by life-style. Your diet, whether you smoke or not, the amount of exercise, and stress can all make a big difference to when and if you suffer these diseases.

We all have to die of something, eventually. However, with a little care in your life-style you can reduce the risk of suffering from some diseases early in your life, and have a better chance of feeling healthy and living for a long time.

13.19 Coronary heart disease and strokes.

Coronary heart disease is often abbreviated to **CHD**. It is a disease which affects the coronary vessels supplying oxygen and nutrients to the heart muscles. In this disease the walls of the coronary arteries are gradually damaged by deposits of lipids and cholesterol. These are laid down in the artery wall. The wall becomes stiff which is why the term 'hardening of the arteries' is sometimes used to describe this disease. The medical term meaning 'hardening of the arteries' is **atherosclerosis**.

Although CHD is called a disease, it is probably a natural process of ageing. However, some people develop it more than others. A number of factors indicate the possible risk of early development of CHD. The most important ones are genetic factors, high blood cholesterol levels and aspects of a person's life-style.

The genetic factor is particularly important. If a person has a father, mother, brother or grandparents who died early of CHD, there is a possibility that the person will also suffer CHD at an early age. If a person has a very high cholesterol level in the blood, that person can be at risk of CHD. Several life-style factors have been linked to CHD, including a high fat diet – especially one rich in saturated fat from animals – lack of exercise, smoking and leading a stressful life.

Once CHD has developed, it is likely that at some point the artery wall cracks on the inside, the platelets get caught on it and a blood clot starts to form inside the artery. Because of the speed of flow of the blood the clot can be swept off the artery wall and down the coronary artery. The artery divides into smaller and smaller arterioles. Eventually, the blood clot gets stuck and blocks the blood vessel. This stops blood from reaching an area of cardiac muscle. The lack of oxygen will stop these muscle cells contracting properly. If the area of muscle affected is big enough the regular beating of the heart is interrupted and we say that the person has suffered a **heart attack**.

Arteries in the brain can also become affected by atherosclerosis. Because their walls are less elastic than they should be the artery is more likely to burst at the peak of pressure when a pulse of blood is expelled from the heart. Blood leaks out of the damaged vessel. The result of this is most severe in the brain where an area of brain tissue can be killed. A blood clot or burst blood vessel in the brain is called a **stroke**, but a blood clot is more likely than a burst blood vessel.

13.20 Regular exercise can reduce the risk of CHD.

In the past twenty years or so, the amount of exercise which many people take has decreased. This is partly because of the kind of work they do. Fewer people now do hard, physical work. More time is spent in offices, doing sedentary jobs. Another reason is the change in the way in which people relax, especially young people. Fewer young people now spend much time doing energetic activities such as sport. They spend more time watching television, listening to music or playing computer games.

Exercise has many beneficial effects on health.

Exercise improves the efficiency of the heart.

Regular exercise improves the strength of the heart muscle. The heart of someone who exercises regularly is likely to be able to pump more blood with each beat than the heart of someone who does not take exercise. The amount of blood pumped in one beat is called the **stroke volume**. An increased stroke volume means that the heart can pump enough blood to the body with fewer beats per minute, so the **resting pulse rate** is lower. Overall, the heart does not have to work so hard. Regular exercise can greatly decrease the risk of suffering from CHD.

Aerobic exercise improves lung capacity.

Exercising muscles need extra oxygen, which is supplied to them by breathing faster and more deeply. If you regularly do exercise which makes your muscles demand extra oxygen, called **aerobic exercise**, this helps your respiratory system to become efficient at getting oxygen into your blood.

Aerobic exercise improves the oxygen-carrying capacity of the blood.

If your muscles constantly demand a lot of oxygen, the number of red blood cells in your blood increases. This makes your blood more efficient at carrying oxygen, so your heart does not have to beat as fast or as hard to supply your muscles.

Exercise improves the strength of your muscles.

The blood supply to muscles increases, so that they are less likely to have to respire anaerobically and produce lactic acid (Section 6.3). The muscles get stronger, and can work for longer without tiring.

Exercise affects the development of bones.

The way in which bones grow partly depends on the forces which act on them, and a person who regularly exercises may end up with stronger bones than a person who does not. Exercise can also make sure that joints stay flexible.

Exercise makes you feel good.

Many people get a lot of pleasure from exercise. Exercise may cause your brain to secrete chemicals which make you feel happy. Exercise also gives you a break from your work, and helps you to feel less stressed and more relaxed.

13.21 Alcohol is a commonly-used drug.

A drug is a substance which changes the way the body works. Many drugs are used in medicine, such as antibiotics and painkillers. Most of us take some drugs every day – for example, caffeine in coffee or tea. However, there are some drugs which can cause great harm if they are not used properly.

Alcohol is a very commonly-used drug in many different countries. People often drink alcoholic drinks because they enjoy the effect that alcohol has on the brain. Alcohol can make people feel more relaxed and release their inhibitions, making it easier for them to enjoy themselves.

Alcohol is quickly absorbed through the wall of the stomach, and carried all over the body in the blood. It is eventually broken down by the liver, but this takes quite a long time.

Drinking fairly small quantities of alcohol is not dangerous, but alcohol does have many effects on the body which can be very dangerous if care is not taken.

Alcohol lengthens reaction time.

Alcohol is a **depressant**, which means that even small amounts of alcohol slow down the actions of parts of the brain, so alcohol lengthens the time you take to respond to a stimulus. This can mean the difference between life and death – often someone else's death – if the affected person is driving a car. A very high proportion of road accidents involve people who have recently drunk alcohol – either drivers or pedestrians. Most countries in which drinking alcohol is allowed have legal limits on blood alcohol level when you drive. However, we now know that even very small quantities of alcohol increase the risk of an accident, so the only safe rule is not to drink alcohol at all if you drive.

Alcohol can increase aggression in some people.

Different people react differently to alcohol. In some people, it increases their feelings of aggression, and releases their inhibitions so that they are more likely to be violent or commit other crimes. They may be violent towards members of their family. Research has shown that at least 50% of violence in the home is related to drunkenness, and that alcohol has played a part in the criminal behaviour of around 60% of people in prison.

Large intakes of alcohol can kill.

Every year, people die as a direct result of drinking a lot of alcohol over a short period of time. Alcohol is a poison. Large intakes of alcohol can result in unconsciousness, coma and even death. Sometimes, death is caused by a person vomiting when unconscious, and then suffocating because their airways are blocked by vomit.

13.22 Alcoholism is a dangerous disease.

Alcoholism is a disease in which a person cannot manage without alcohol. The cause of the disease is not fully understood. Although it is obvious that you cannot become an alcoholic if you never drink alcohol, many people regularly drink large quantities of alcohol, but do not become alcoholics. Probably, there are many factors which decide whether or not a person becomes alcoholic. They may include a person's genes, their personality, and the amount of stress in their lives.

An alcoholic needs to drink quite large quantities of alcohol regularly. This causes many parts of the body to be damaged, because alcohol is poisonous to cells. The **liver** is often damaged, because it is the liver which has the job of breaking down alcohol in the body. One form of liver disease resulting from alcohol damage is **cirrhosis**, where fibres grow in the liver. This can be fatal.

Excessive alcohol drinking also damages the **brain**. Over a long period of time, it can cause loss of memory and confusion. One way in which the damage is done is that alcohol in the body fluids draws water out of cells by osmosis. When this happens to brain cells, they shrink, and may be irreversibly damaged. This osmotic effect is made worse because alcohol inhibits the release of a hormone which stops the kidneys from allowing too much water to leave the body in the urine. So drinking alcohol causes a lot of dilute urine to be produced, resulting in low levels of water in the blood.

13.23 Heroin is an addictive narcotic.

Opium poppies produce a substance called opium, which contains a number of different chemicals. Some of these, especially morphine and codeine, are used in medicine for the relief of pain. Opium is also the raw

13.13 Alcohol abuse can cause misery to the abuser and to their friends and family.

material from which heroin is produced, but heroin is used only very rarely in medicine, because it is very addictive.

Heroin, like alcohol, is a powerful depressant, slowing down many functions of the brain. It reduces pain, slows down breathing and the functions of the hypothalamus. When a person takes heroin, it produces a feeling of euphoria – that is, they feel intensely happy. However, in many people it can rapidly become addictive. They feel so ill when they do not take it that they will do anything to obtain more. As their bodies become more tolerant of the drug, they need to take more and more of it in order to obtain any feelings of pleasure.

Not everyone who takes heroin becomes addicted to it, but many do. Addiction can develop very rapidly, so that a person who has taken it for only one or two weeks may find that they cannot give it up.

A person who has become addicted to heroin may lose any ability to be a part of normal society. He or she may think only of how they will get their next dose. They may not be able to hold down a job, and cannot earn money, so many heroin addicts turn to crime in order to obtain money to buy their drug. They are not able to help and support their family.

Some people take heroin by injecting into their veins. This can be dangerous as the needles used for injection are often not sterile, and pathogens such as the hepatitis virus can be introduced into the body. The sharing of needles by heroin addicts has been a major method by which HIV has spread from person to another.

It is possible for a heroin addict to win the battle against his or her addiction, but it needs a great deal of will-power and much help from others. The withdrawal symptoms that an addict suffers after a few hours without the drug can be extremely unpleasant, and even life threatening.

Chapter revision questions

1. For each of the following types of disease
   (i) name one example,
   (ii) briefly describe the cause of your example, and
   (iii) briefly describe the symptoms of your example.
   (You may have to use the index to find information for some of your answers to part (iii).)
   (a) a deficiency disease
   (b) an inherited disease
   (c) a degenerative disease
   (d) an infectious disease

2. Explain why:
   (a) you can normally get measles only once, but you can get a cold over and over again
   (b) babies fed on breast milk are less likely to get an infectious disease than babies fed on bottled milk
   (c) transplants are more likely to be successful if the donated organ comes from a close relative of the recipient
   (d) it is no use trying to treat a cold with antibiotics.

3. Discuss the ways in which life-style can affect a person's health.

4. Explain the difference between each of the following terms.
   (a) pathogen and antigen
   (b) antibody and antibiotic
   (c) heart attack and stroke
   (d) bacteria and viruses
   (e) phagocyte and lymphocyte
   (f) AIDS and HIV

5. (a) Describe the signs, symptoms and effects of the disease syphilis. [6]
   (b) Explain
      (i) how HIV is transmitted, and
      (ii) how its spread can be prevented
   (c) Explain why the methods for treating syphilis cannot be used for the treatment of AIDS. [2]

   0610/3 N98

6. (a) What type of organism causes each of the following?
   AIDS
   Gonorrhoea
   Syphilis [3]
   (b) Which of the above cannot be successfully treated with antibiotics? [1]

   0610/3 J96 (modified)

Chapter revision questions (continued)

7 Fig 13.14 shows a section through a diseased blood vessel of a middle-aged person.

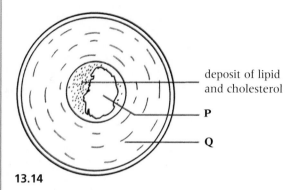

13.14

(a) What type of blood vessel is shown? [1]
(b) Name the parts labelled **P** and **Q**. [1]
(c) Describe the possible effects of the deposit of lipid and cholesterol on the person's health. [4]
(d) Explain how the person's diet, over the previous twenty years, may have caused this blood vessel to be diseased. [2]
5090/2 J00 (modified)

8 Urine samples were taken from three different people, **L**, **M** and **N**. Each sample was tested for the presence of alcohol, glucose and protein. The table below shows the results obtained.

person	alcohol	glucose	protein
L	✓	✗	✓
M	✗	✓	✗
N	✓	✗	✗

key: ✓ = present ✗ = absent

(a) (i) Which person is likely to be producing insufficient insulin? [1]
(ii) Explain your answer. [2]
(b) (i) Suggest which person might be suffering from kidney disease. [1]
(ii) Explain your answer. [3]
0610/3 J96 (modified)

9 Fig 13.15 shows a section through a virus.

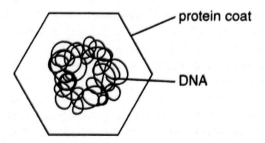

13.15

(a) State **two** differences between this virus and a bacterium. [2]
(b) (i) Name the virus that causes AIDS. [1]
(ii) List **three** ways in which the AIDS virus may be spread. [3]
5096/1 J00

10 Fig 13.16 shows an organism **W** and 13.17 shows how the reproduction of this organism is affected by an antibiotic.

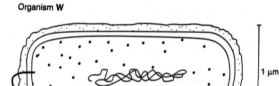

13.16

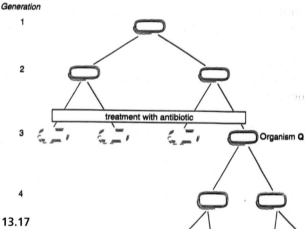

13.17

(a) (i) What type of organism is **W** most likely to be? [1]
(ii) State **three** reasons for your answer. [3]
(b) Name the type of reproduction shown by organism **W**. [1]
Q is the only organism surviving the antibiotic treatment.
(c) Suggest an explanation for the survival of **Q** and its offspring. [2]
(d) Explain why patients who are treated with antibiotics are always advised to take the complete course of treatment, rather than stop the treatment as soon as they feel better. [3]
0610/3 J98

14 Making use of microorganisms

14.1 Biotechnology is not new.

Today, many industrial processes use **biotechnology**. Biotechnology means using living organisms to carry out processes which make substances which we want. The term is normally only used when *micro*organisms are used, or when plants or animals are used to produce something other than food. So most people would not include farming animals and crops for food as biotechnology.

Although the word 'biotechnology' only began to be widely used in the 1970s, we have been using organisms to make things for us for thousands of years. Yeast has been used to make alcoholic drinks, and to make bread, yoghurt and cheese. Bacteria and other organisms have been used to make compost. More recently, we have learned how to use bacteria and protoctists to make waste substances such as sewage harmless. Useful fuels such as biogas may be made by these processes, too.

Even more recently, in the last twenty years or so, genetic engineering has opened up a whole new world of possibilities for biotechnology. With genetic engineering, we can introduce almost any gene we like into microorganisms or other living things. The organisms may then make the protein for which the gene codes. This allows us to grow bacteria which make human hormones, or plants which produce drugs, for example.

14.2 Microorganisms include viruses, bacteria, some fungi and protoctists.

A microorganism is any organism which is too small to be seen without a microscope. Fig 14.1 shows examples of the four groups of microorganisms. You can find a summary of the characteristic features of bacteria and fungi in Chapter 17.

As you probably know, some microorganisms are harmful. Bacteria and fungi may make food go bad. All four of these groups may cause diseases in humans or in crop plants. This is described in Chapter 13.

However, many microorganisms are very useful to us. Bacteria and fungi, in particular, have essential roles to play in the carbon and nitrogen cycles where some of them act as **decomposers**, breaking down the remains of dead animals and plants. You can read more about this in Sections 15.10–15.14. In this chapter, we will look at many other ways, both old and new, in which we make use of microorganisms.

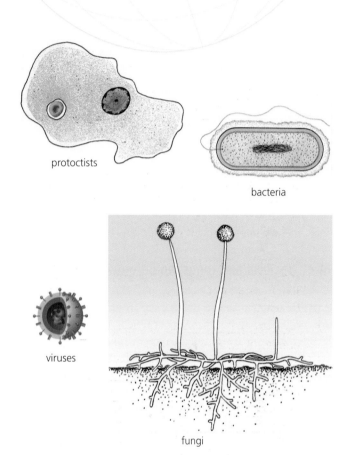

14.1 The four groups of microorganisms.

Growing microorganisms

14.3 Viruses can only be grown in living cells.

You will not be able to do any experiments with viruses. This is because viruses are very difficult to grow. Viruses can only be grown inside living cells.

Fig 14.2 (overleaf) shows the life cycle of a virus. You can see that it reproduces only when it is inside another living cell. This particular virus reproduces inside a bacterium. Even bacteria can get ill!

Scientists do sometimes need to grow viruses. They may need, for example, to grow viruses which they have taken from someone who is ill, to find out exactly what sort of virus it is. They normally do this by injecting the viruses into hen's eggZ, or into a culture of human cells. It takes quite a long time for the viruses to reproduce enough for the scientists to get a large sample of them.

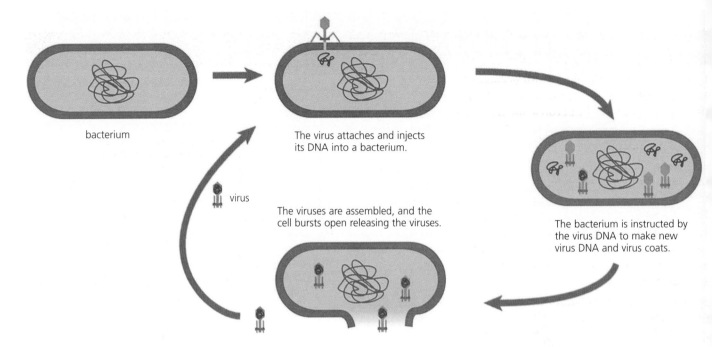

14.2 The life cycle of a virus which infects bacteria.

14.4 Bacteria and fungi can be grown on agar jelly.

Bacteria and fungi are quite easy to grow. You can try one way of growing them in Practical 14.1.

Bacteria and fungi need three things – water, a food source and a warm temperature – if they are to grow and reproduce quickly. Most of them also need oxygen. If we want to grow a large culture of bacteria or fungi, we need to give them good supplies of these four things. When these microorganisms are grown on a large scale, they are often grown in a **fermenter**, which is designed to supply all of their needs. You can see one example in Fig 14.11.

If we want to *stop* them growing, then we need to deprive them of one or more of these things. We do this when we preserve food.

Practical 14.1 Growing bacteria and fungi on agar jelly

In this practical, you are going to grow microorganisms from two different sources, pond water and tap water.

You need to keep all of your apparatus **sterile**. This means that no living microorganisms must be on it. This is because you want to be sure that any bacteria or fungi which you grow have come from the pond water or the tap water.

You will grow the microorganisms on agar jelly. This is made from seaweed. It is bought as a powder, which is mixed with water and then heated to boiling point. This kills any living organisms. The mixture is then allowed to cool to just below 60 °C. It is then poured into a sterile petri dish. It sets in the dish to form a layer of jelly, containing nutrients and water for bacteria and fungi to feed on.

The petri dishes you will use will probably be made of plastic. These are sterilised by the manufacturers (you cannot sterilise them in school by heating them, because they would melt!).

When you handle them, you must make sure that you do not allow any microorganisms to get *inside*, except the ones in the pond water or tap water. For example, when the hot agar jelly is poured in, the top of the dish must only be lifted just a little way, and the jelly poured in quickly. Otherwise, microorganisms might drop in from the air.

1. Collect three petri dishes containing sterile agar jelly. Just by handling the dishes, you will get microorganisms on the outside of them - but this does not matter, so long as you do not let any microorganisms get *inside*.

   Use a glass marking pen to label the dishes by writing on the lid. Label one of them 'pond water', another one 'tap water', and the third one 'control'. Write your name on the dishes, too.

2. Using adhesive tape, tape down the lid of the 'control' dish as shown in Fig 14.3.

cal 14.1 (continued)

his is done so that the lid cannot accidentally be moved, which could let microorganisms in. Don't pe all around the dish, because air needs to be able get in.

ollect a small container of pond water, and an **oculating loop**. This is a piece of looped wire tached to a handle. You will use it to spread some nd water onto the agar jelly in the 'pond water' dish. Before you do this, you need to get rid of any icroorganisms on the loop. Fig 14.3 shows you how do this, and how to spread some pond water onto e agar jelly. When you have done this, tape down e lid as before.

epeat step 3 using tap water instead of pond water, d spreading it onto the petri dish labelled tap water.

w put the three petri dishes into an incubator at out 25°C. This temperature is warm enough to let y microorganisms grow quite quickly, but not warm ough to encourage the growth of the sort of icroorganisms which might make you ill. (At what mperature do you think such pathogenic icroorganisms would grow best?) Put your petri shes upside down, so that any condensation in the sh does not fall onto the surface of the agar jelly.

ter a day or so, look at your three dishes to see if ything has grown on them, and make labelled agrams of each of them. Fig 14.4 shows what a dish ight look like. You may see **colonies** of bacteria and ngi on the surface of the agar. Although a single cterium is much too small to see, if one breeds and rms a colony of many thousand, then they become sible. Each colony on your dish probably grew from st a single microorganism, which was in the water hich you spread on the agar.

tions

part from stopping microorganisms from getting in, n you suggest another reason for taping down the ds of the petri dishes?

hat was the purpose of the control petri dish?

as there any difference in the numbers or kinds of icroorganisms which came from the pond water and e tap water?

u have no idea what sort of microorganisms you ve grown on your agar jelly. Some of them might be rmful to humans. What do you think should be ne with all of the petri dishes once you have ished with them?

1 Pouring the agar plate

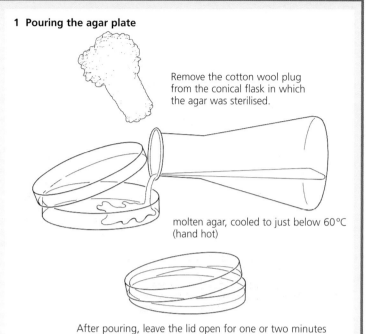

Remove the cotton wool plug from the conical flask in which the agar was sterilised.

molten agar, cooled to just below 60°C (hand hot)

After pouring, leave the lid open for one or two minutes to release steam, and then gently tip the lid down.

2 Using the inoculating loop

First dry the loop by holding it in hot air above the Bunsen flame (not in it). Then flame the loop by putting it into the flame at an angle. Let the wire loop get red hot.

Dip the loop into your pond water.

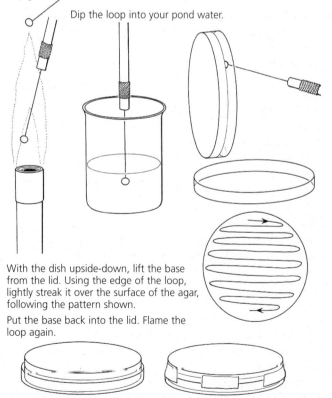

With the dish upside-down, lift the base from the lid. Using the edge of the loop, lightly streak it over the surface of the agar, following the pattern shown.

Put the base back into the lid. Flame the loop again.

Securely tape the lid to the base. Incubate upside-down.

14.3 Growing bacteria on agar.

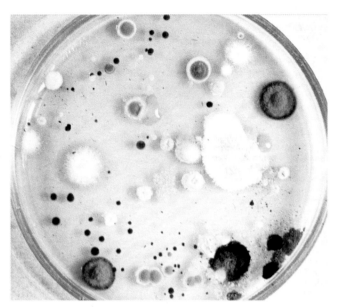

14.4 Bacteria and fungi growing on agar in a petri dish. The 'furry' colonies are fungi and the smoother ones bacteria. Each colony grew from a single cell.

Sewage treatment

14.5 Sewage can be harmful to people and the environment.

Sewage is waste liquid which has come from houses, industry and other parts of villages, towns and cities. Some of it has just run off streets into drains when it rains. Some of it has come from bathrooms and kitchens in people's houses and offices. Some of it has come from factories. Sewage is mostly water, but also contains many other substances. These include urine and faeces, toilet paper, detergents, oil and many other chemicals.

Sewage should not be allowed to run into rivers or the sea before it has been treated. This is because it can harm people and the environment. Untreated sewage is called **raw sewage**.

Raw sewage contains many bacteria and other microorganisms, some of which may be pathogenic. Some examples of these are described in Chapter 13. People who come into contact with raw sewage, especially if it gets into their mouths, may get ill.

Raw sewage contains many substances which provide nutrients for plants and microorganisms. These can cause **eutrophication**. This is described in Section 16.13.

It is therefore very important that sewage is treated to remove any pathogenic organisms, and most of the nutrients, before it is released as **effluent**. Microorganisms play an important part in all the most commonly used methods of sewage treatment.

When sewage has been treated, the water in it can be used again, so sewage treatment enables water to be recycled. It may not be a nice thought to know that the water you drink was once inside someone else's body, but if we did not recycle water in this way then significant water shortages would occur in many parts of the world.

14.6 Liquids from sewage can be treated by two different methods.

Fig 14.5 shows how sewage is treated to make it safe. First, the raw sewage is passed through **screens**. These trap large objects such as grit which may have been washed off roads. The screened liquid is then left for a while in **settlement tanks**, where any other insoluble particles drift to the bottom and form a sediment.

There are two different ways in which the resulting liquid can now be treated.

Trickling filters

The liquid from the settlement tanks is sprinkled over a trickling filter bed. This is made of small stones and clinker. Many different aerobic microorganisms live on the surface of the stones. Some of them are **aerobic bacteria**, which feed on various nutrients in the sewage. **Protoctists** feed on the bacteria. **Fungi** feed saprophytically on soluble nutrients. These microorganisms make up a complex ecosystem in the trickling filter bed.

The liquid is trickled onto the surface of the stones through holes in a rotating pipe. This makes sure that air gets mixed in with the liquid. The liquid trickles quite slowly through the stones, giving the microorganisms plenty of time to work on it. By the time the water drains out of the bottom of the bed, it looks clear, smells clean, contains virtually no pathogenic organisms, and can safely be allowed to run into a river or the sea.

Activated sludge

In this method, the liquid from the settlement tanks runs into a tank called an **aeration tank**. Like the trickling filter bed, this contains **aerobic microorganisms**, mostly bacteria and protoctists. Oxygen is provided by bubbling air through the tank. As in the trickling filter bed, these aerobic microorganisms make the sewage harmless.

Why is this method called 'activated sludge'? 'Activated' means that microorganisms are present. Some of the liquid from the tank, containing these microorganisms, is kept to add to the next lot of sewage coming in. 'Sludge' means just what it sounds like! It is a word which describes the semi-solid waste materials in sewage.

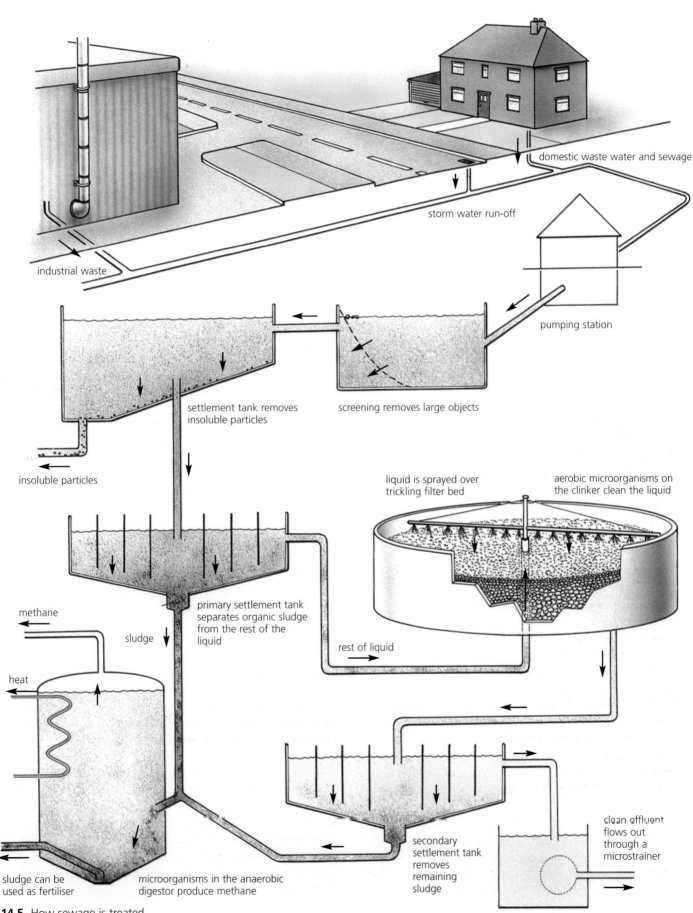

14.5 How sewage is treated.

151

Both the trickling filter and the activated sludge methods can run into problems if the sewage contains substances which harm the microorganisms. These include heavy metals such as mercury, disinfectants, or large quantities of detergents. Heavy metals and disinfectants are toxic to many of the microorganisms. Detergents may cause foaming, which stops oxygen getting into the liquid. To solve these problems, the contaminated sewage can be diluted before being allowed to enter the trickling filter bed or the activated sludge tank.

14.7 Sludge can be digested anaerobically.

So far, we have described how the *liquid* part of the sewage is treated. What about the *solid* part?

Solids – sludge – first dropped out of the sewage in the settlement tank. The activated sludge method also produces sludge. This material contains lots of living and dead microorganisms. It contains valuable organic material. It is a pity to waste it.

The sludge can be acted on by **anaerobic bacteria**. The sludge is put into large, closed tanks. Inside the tanks, several different kinds of bacteria act on the sludge. Some of them produce **methane**, which can be used as a fuel. When they have finished, the remaining solid material has to be removed from the tank. It is often used as fertiliser – it is usually quite safe, because it is very unlikely that any pathogenic organisms will have survived all these processes (Fig 14.5).

Making food with microorganisms

14.8 Yeast is used for making alcohol.

Yeast is a single-celled fungus. Fig 14.6a shows the structure of a yeast cell.

Yeast cells feed **saprotrophically**. This means they secrete enzymes from their cells. The enzymes digest the food on which the yeast is living, breaking down large molecules into small ones. The small molecules then diffuse into the yeast cell.

'Wild' yeast grows in many different places. It usually grows on foods which contain sugar, such as fruit. When we grow yeast, we need to provide it with the types of food which it needs. It is usually grown in a solution containing carbohydrate – usually in the form of sugar – and minerals, including ammonium ions. Each yeast cell absorbs the sugar and minerals, and uses some of them to grow. When the cell gets to a certain size, it produces a new cell by budding (Fig 14.6b). Yeast cells reproduce fastest when the temperature is quite warm, around 40 °C.

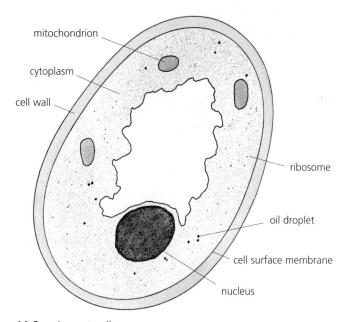

14.6a A yeast cell.

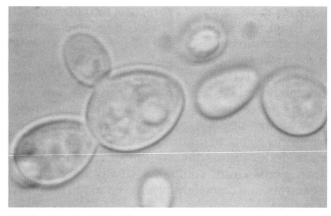

14.6b Yeast cells budding.

Questions

1. What is biotechnology?
2. List the four groups of microorganisms.
3. Explain why you cannot grow viruses in your school laboratory.
4. What do bacteria and fungi need, in order to grow well?
5. Sewage is more likely to need diluting before treatment if it comes from industrial areas than if it comes from rural areas. Why do you think this is so?
6. Suggest an advantage of the activated sludge method, compared with the trickling filter method, of sewage treatment.
7. Why is the tank used for digestion of the sludge closed, rather than with an open top?

People have been using yeast for thousands of years to make **alcohol**. If yeast is added to a sugar solution, it absorbs some of the sugar into its cells. It then uses this sugar in respiration.

Usually, the yeast respires **anaerobically**. When it does this, it converts the sugar to ethanol (a type of alcohol) and carbon dioxide. This process is called **alcoholic fermentation**. 'Fermentation' is a name for any type of respiration which makes a product other than carbon dioxide and water. The equation for alcoholic fermentation is:

glucose $\longrightarrow$ alcohol + carbon dioxide + energy

Many different alcoholic drinks are made in this way. **Beer** is made by providing the yeast with the sugar **maltose**, obtained from germinating barley seeds. Usually, hops are added as well, to give a bitter flavour to the beer. **Wine** is made by providing the yeast with sugar from grapes.

Another use for the alcohol is as a **fuel**. In Brazil in the 1980s and early 1990s, sugar cane was grown especially for making alcohol. Yeast was grown in sugar solutions made from the cane. The alcohol produced was purified by distillation, and used to fuel motor vehicles. This provides a renewable source of fuel, rather than using fossil fuels which are non-renewable. The alcohol also causes less pollution than fossil fuels, because it does not produce sulphur dioxide and nitrogen oxides when it burns. However, it is an expensive process. In 1994, world fossil fuel prices dropped, making it more economic for Brazil to import fossil fuels rather than to make its own alcohol in this way.

14.9 Yeast is used for making bread.

When yeast respires, it produces carbon dioxide. If this happens inside a dough made from flour and water, the bubbles of carbon dioxide get trapped in the dough, and make it rise. This is how bread is made.

The dough is made using flour made from cereal grains, usually wheat. The flour contains starch, amylase and protein.

The **starch** is the energy source for the yeast. The **amylase** digests the starch to sugar, so that the yeast can absorb it and use it in respiration. Some bread flours have extra amylase added to them to speed up this process. The amylase does not begin acting on the starch until water is mixed with the flour.

The **protein** is important for the texture of the bread. The most important protein in bread flour (apart from the amylase!) is called **gluten**. This forms sticky, stretchy threads as the yeast works on the dough. This helps to trap the bubbles of carbon dioxide, and makes the dough rise well. Farmers who want to sell their wheat for bread making choose varieties which are known to produce grains containing a lot of gluten.

To make bread, yeast, sugar, flour and water are mixed together to make a dough, which is then left in a warm place to rise. The dough is then mixed again, and made into the shapes of the loaves to be made. It is left to rise again, and then baked. The high temperatures kill the yeast, break down the alcohol which it has made, and alter the remaining starch and gluten to make a firm textured bread.

> ### Practical 14.2 What factors affect the way that dough rises?
>
> You can make a 'basic' bread dough in the following way.
>
> 1 Mix about 0.5 g of sugar into about 50 cm^3 of warm water.
> 2 Mix about 1 g of dried yeast into the warm sugar solution, and leave for a few minutes.
> 3 Measure out about 75 g of flour.
> 4 Add the yeast and sugar mixture to the flour, and pull it around with your hands to make a dough.
> 5 Leave the dough to rise in a warm place covered with cling film.
>
> There are many factors which affect *how much* the dough rises, and also *how fast* it rises. Choose two factors from the following list to investigate.
>   (a) temperature
>   (b) ratio of yeast to flour
>   (c) addition of salt (sodium chloride)
>   (d) type of flour used
>   (e) type of yeast used
>   (f) addition of 'flour improvers', such as ascorbic acid (vitamin C)
>   (g) addition of amylase
>
> If you are able to do this practical in a room used for food preparation, then you could bake and eat your bread. However, if you are working in a science laboratory, then it is not safe to do this.

14.7 Carbon dioxide produced by respiring yeast caused this bread to rise.

14.10 Bacteria are used to make yoghurt and cheese.

Some bacteria, like yeast, respire anaerobically when provided with a suitable source of sugar. One bacterium which does this is called *Lactobacillus*.

Lactobacillus uses sugar from milk as its energy source. This sugar is called **lactose**. *Lactobacillus* converts the lactose to **lactic acid**.

$$\text{lactose} \longrightarrow \text{lactic acid} + \text{energy}$$

Lactic acid, like all acids, tastes sour. However, most people like its taste. The presence of the lactic acid lowers the pH of the milk. This affects the proteins in the milk. They **coagulate**, forming clumps. The milk separates out into these clumps, called **curds**, and a liquid, called **whey**.

Yoghurt is made using a species called *Lactobacillus bulgaricus*. A culture of the bacterium is simply added to warm milk, and left for a few hours. Usually, the milk is heated to around 70 °C, and then cooled, before the *L. bulgaricus* is added. This is to kill any other microorganisms in the milk, which might also ferment it, making different, unpleasant-tasting substances.

Other species of *Lactobacillus* are used to make **cheese**. Sometimes, an enzyme called **rennin** is added to the milk along with the *Lactobacillus*. This enzyme acts on the protein in the milk, making it coagulate even more than it would with just the bacterium. The curds and whey are then separated. The whey is used for making sweets, or for animal feeds, or sometimes just thrown away. The curds are pressed and made into cheese. Different kinds of cheeses are made by using different sorts of milk, different mixtures of bacteria, letting the bacteria work at different temperatures, adding different amounts of salt, pressing the curds or leaving them soft, and leaving the cheese to ripen for different lengths of time or in different conditions. No wonder there are so many kinds of cheese!

In some cheeses, **fungi** are added as well. The blue streaks in blue cheeses such as blue Stilton or Roquefort are fungal hyphae. Sometimes, the fungal spores just fall in from the surroundings. This is especially likely if blue cheeses have been made in the same place before – the fungal spores from previous cheeses will be floating around in the air. Usually, though, the spores are intentionally added to the cheese. You can sometimes see that the blue streaks are in straight lines. This is because a thin wire was coated with fungal spores, and then pulled through the curds while they were setting. The fungi need oxygen, so the cheese-maker will also need to make small holes in the cheese to allow air to get into it.

Large holes in cheeses such as Emmenthal are bubbles of carbon dioxide, produced by bacteria.

Questions

1 Which of the following processes use **anaerobic fermentation**, and which use **aerobic fermentation**?
   (a) making alcohol from sugar
   (b) making lactic acid from lactose

2 When alcohol or lactic acid are being made by microorganisms, the mixture gets hot. Why?

Practical 14.3 Making yoghurt

Here is a basic yoghurt recipe. You will need to make sure that all the apparatus you use has been sterilised. If not, then other microorganisms will act on the milk, producing substances which you do not want.

1 Collect some 'live' yoghurt. This is yoghurt which has not been heat treated, so it still contains living *Lactobacillus*. This is your 'starter culture'. You will need roughly 1 cm³ of starter culture for each sample of yoghurt which you make.

2 Measure 10 cm³ of milk into a sterile container, such as a test tube.

3 Add the starter culture, and mix gently. Cover the tube with cling film, to stop any other microorganisms getting in.

4 Stand your tube in a water bath or incubator at about 40 °C, and leave for approximately two hours for the bacteria to turn the milk into yoghurt.

Many factors affect the speed at which the yoghurt is formed, and the kind of yoghurt which is made. You can tell what is happening just by looking at your milk/yoghurt, or you could test its pH (because the bacteria are producing lactic acid). Choose two of the following factors, and investigate how they affect the rate of action of the bacteria, and/or the final properties of the yoghurt which is produced.
   (a) the type of milk used
   (b) the type of starter culture used
   (c) whether air can get to the milk or not
   (d) the temperature at which the milk is kept
   (e) adding lactase to the milk

If you are able to do this practical in a room used for food preparation, then you could taste your yoghurt. However, if you are working in a science laboratory, then it is not safe to do this. You should not eat your yoghurt if you have added anything to your milk and starter culture mixture, such as enzymes, just in case they make you ill.

14.8 In the container are curds, which are being made into cheese. The acidity is being checked. What makes it acid?

14.11 Microorganisms can be used as food.

In the production of alcohol, bread, yoghurt and cheese, we use microorganisms to change one substance into another, which we use as food. But we can also use the microorganisms themselves as food.

There are many good arguments for doing this. In many parts of the world, there is a shortage of food, especially protein-rich food. Microorganisms could provide a good source of protein in these areas. Microorganisms do not need soil to grow in. They can use many different substances as food sources, including wastes from other processes – so they could be grown very cheaply. Producing food in this way wastes less energy than producing meat, because it 'taps in' to an earlier stage in the food chain (Section 15.9).

The first attempts to make microorganisms into food used yeast. In Germany during World War I, yeast was cultured in large vats, using molasses as a food source for the yeast, to produce a protein supplement for people. More recently, different kinds of protoctists (usually single-celled photosynthetic ones) and bacteria have been grown for food production. The food made from all of these microorganisms is known as **single cell protein** or **SCP**.

14.9 Mycoprotein can be made to look like pieces of meat. It is quite bland and can be cooked with other ingredients.

However, there have been big problems in selling SCP as food for people. People are very suspicious of eating microorganisms – even though they like eating yoghurt and cheese! The first SCPs also tasted rather unpleasant, partly because they contained a lot of DNA and RNA, which tastes bitter. Most SCPs are now marketed as animal feed.

One SCP which has found a market as human food, however, is **mycoprotein**. This is made from a fungus. In Britain, the fungus which is used is called *Fusarium*. Its structure is rather like that of *Mucor* (Fig 17.4), so it is made of hyphae rather than single cells. But mycoprotein is still often called SCP!

The *Fusarium* is grown in large vats, using carbohydrates as a food source, with other nutrients such as ammonium nitrate added as well. The carbohydrates often come from waste left over from making flour. *Fusarium* reproduces quickly and makes a mass of mycelium, which is harvested and treated to remove a lot of the RNA which it contains. Then it is dried, and shaped into chunks or cakes, ready for eating as it is, or for making into pies or other foods. Some people think that mycoprotein looks and tastes a bit like chicken. If you have not seen any, have a look for some next time you are shopping in a supermarket. In Britain it is sold as Quorn®.

Mycoprotein is an excellent food. It has a high protein content, very little fat, no cholesterol, and a lot of fibre. Because the mycelium of the fungus is made up of long thread-like hyphae, mycoprotein has a fibrous texture which many people like – because it is a bit like meat. It has quite a bland taste, and can easily be flavoured to make a pleasant-tasting food.

Microorganisms and medicine

14.12 Fungi make antibiotics.

Antibiotics are substances which kill bacteria without harming human cells. We take antibiotics to help to cure bacterial infections. You can read about how one of them – penicillin – works, and some of the problems involved with the use of antibiotics, in Section 13.17.

Penicillin is made by growing the fungus *Penicillium* (Fig 14.10) in a large fermenter (Fig 14.11). It is grown in a culture medium containing carbohydrates and amino acids. The contents of the fermenter look a bit like porridge. They are stirred continuously. This not only keeps the fungus in contact with fresh supplies of nutrients, and mixes oxygen into the culture, but also rolls the fungus up into little pellets. This makes it quite easy to separate the liquid part of the culture – which contains the pencillin – from the fungus, at a later stage.

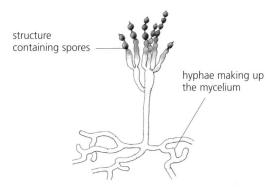

14.10 *Penicillium*, the fungus that makes penicillin.

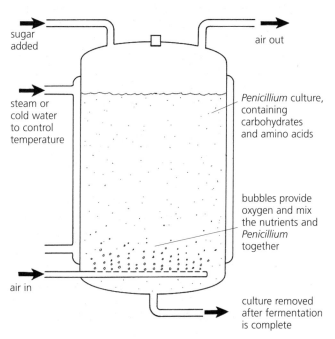

14.11 A fermenter used for producing penicillin.

To begin with, the fungus just grows. This stage takes about 15–24 hours. After that, it begins to secrete penicillin. The rate at which it produces penicillin partly depends on how much sugar it has available. If there is a lot of sugar, then not much penicillin is made. If there is no sugar at all, then no penicillin is made. So small amounts of sugar have to be fed into the fermenter all the time that the fungus is producing penicillin.

The culture is kept going until it is decided that the rate of penicillin production has slowed down so much that it is not worth waiting any longer. This is often after about a week, although the exact time can vary quite a lot on either side of this. Then the culture is filtered, and the liquid treated to concentrate the penicillin which it contains.

> **Question**
>
> Like penicillin, single cell protein is made in a fermenter. However, when mycoprotein is made, there is no stirrer in the fermenter. Suggest why this is so.

Genetic engineering

14.13 Genes can be transferred from one organism to another.

A gene is a length of DNA which codes for the production of a particular protein by a cell. We are now able to take genes from one organism and put them into another. This is called **genetic engineering**.

To explain how this is done, we will look at the way in which genetic engineering is used to produce insulin.

14.14 Human insulin genes are inserted into bacteria.

Some people are not able to make the protein **insulin**. Insulin is a hormone which helps to regulate the concentration of glucose in your blood. People whose bodies cannot make insulin have the disease **diabetes mellitus**. They have to have injections of insulin every day.

For a long time, the only source of insulin was from animals which had been killed for food, such as pigs. Now, genetic engineering has produced bacteria which make human insulin.

The process begins with the extraction of the gene for making insulin from human cells. This is done using enzymes which chop up DNA molecules into short lengths. The particular length of DNA which codes for making insulin is identified, and separated from all the unwanted DNA.

Now the DNA carrying the gene for insulin must be inserted into a bacterium. This is not easy – you cannot just suck up some DNA with a syringe and inject it into a bacterial cell. One way of getting DNA into a bacterium is to use a **plasmid**. A plasmid is a ring of DNA, found in bacteria, which is able to reproduce itself inside other living cells.

First, some of the DNA in the plasmid is cut out, using enzymes like the ones used for cutting up the human DNA. This leaves a gap in the ring. The human DNA is then mixed up with the plasmid, and a different kind of enzyme added. This enzyme sticks DNA together. It sticks the human DNA into the gap in the plasmid. (You can think of the first kind of enzymes acting like scissors, and this second kind as acting like glue.)

Next, these genetically engineered plasmids are added to a culture of a bacterium. A few of the bacteria will take up one or more plasmids into their cells. These bacteria now contain the human insulin gene. They can be grown in large vats where they secrete insulin into the culture solution.

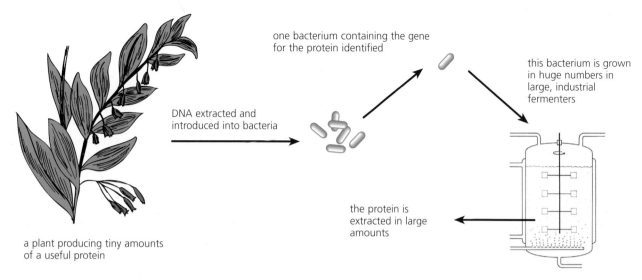

14.12 How genetic engineering can be used to produce a protein in large quantities.

14.15 Many other proteins can be made by genetic engineering.

Several different kinds of proteins are now made by genetically engineered bacteria.

Human growth hormone

Human growth hormone is now made in this way. This is a hormone normally made by the pituitary gland, which has a wide range of roles in the human body. One of them is, as its name suggests, controlling growth. Children who do not secrete enough growth hormone do not grow properly. For many years, the only treatment for them was injections of growth hormone taken from the pituitary glands of dead people. Not surprisingly, the hormone was in very short supply, and very expensive. There was also the risk of diseases, such as the brain disease Creutzfeld-Jakob disease, being spread from the dead person into the child being given the treatment. Now that genetically engineered human growth hormone is widely available, it is much cheaper and much safer to use.

Factor VIII

Factor VIII is also made by genetic engineering. This is a protein normally present in the blood, which helps in clotting. Some people lack the gene which codes for the production of this protein, so their blood cannot clot properly. They have the disease **haemophilia**. Until genetic engineering was invented, factor VIII was extracted from donated blood. It was given regularly to people with haemophilia. In the 1970s, when AIDS first appeared and before it was known that it was caused by HIV, many people with haemophilia were given blood containing HIV. They later developed AIDS. (There is no risk of this happening now, because all donated blood is screened to make sure that it does not contain HIV.)

Genetic engineering is also used to produce enzymes used in food manufacture. Enzymes, of course, are proteins, so genes code directly for their production. You may remember that, in cheese making, an enzyme called **rennin** is added to clot the milk. This enzyme was often extracted from the stomachs of calves. Not surprisingly, vegetarians do not want to eat cheese which been made using this enzyme! However, rennin can now be made by genetically engineered bacteria. Most vegetarians are happy to eat foods such as vegetarian cheese, which have been made using this kind of rennin.

14.16 The characteristics of animals and plants can be changed by genetic engineering.

Genetic engineering does not always involve inserting genes into bacteria. It is possible to insert genes into the cells of almost any organism at all.

For example, in tomatoes, there is a gene which controls ripening. Tomatoes with a particular variety of this gene do not go bad so quickly. This makes it easier to transport them from the grower to the shop, and they can be kept on the shelves for longer – they have a longer shelf life. A new variety of tomato has been produced, by genetic engineering, which contains this gene.

Another possible use of genetic engineering in crop plants is to insert genes which give resistance to herbicides (chemicals used to kill weeds). Farmers could then spray their crops with the herbicide, which would kill the weeds and not the crop.

If you look at newspapers and magazines, you will probably find many more examples of genetic engineering being used to produce new varieties of crop plants, because a great deal of money and research is going on in this area at the moment.

We can also insert genes into animal cells, including human cells. Although research is still in its early stages, there are hopes that we may be able to use a technique called **gene therapy** to help people who have genetic diseases like cystic fibrosis. There have been some fairly successful experiments with this so far.

People with cystic fibrosis are lacking a gene needed for making normal mucus in their lungs. This gene has been extracted from human cells, and inserted into a virus. The virus is one which can infect human cells, but it has been damaged in such a way that it should not make you ill. The viruses, containing the human gene, have then been inserted into the air passages of volunteers with cystic fibrosis. The idea is that the viruses – as viruses do – will inject their DNA into the cells lining the respiratory passages. That, of course, will include the human gene which the volunteers are lacking. There are, however, still several problems with this technique, such as the viruses still causing mild illness, or not enough genes getting into the human cells.

14.17 There are social, ethical and moral concerns about genetic engineering.

Genetic engineering, as you have seen, has opened up many possibilities of improving people's lives. It has also opened up many possible dangers.

For example, genetic engineering means producing bacteria and viruses with different genes from their usual ones. This could cause **health hazards**. What if some of these microorganisms were changed in such a way that they became pathogenic? They might cause a new disease, for which there is no cure. To try to make sure that this does not happen, strict regulations are enforced about the kinds of microorganisms which can be used in genetic engineering, the kinds of genes which can be put into them, and the conditions under which they can be kept. However, what if someone wanted *deliberately* to produce a new pathogen, and release it into the environment?

As well as health hazards, genetic engineering could produce **environmental hazards**. For example, imagine that a new variety of rape is produced, which has been genetically engineered to be resistant to a particular insect which feeds on it. This gene might be passed – perhaps in pollen – to closely related, wild, plants growing nearby. The gene might spread through the population of wild plants. This could upset the food web in the ecosystem, because insects could no longer feed on the wild plants. So far, there have been no instances of this happening, and it is thought to be very unlikely – but it *is* just possible.

Genetic engineering also raises issues about what is morally and ethically acceptable to society. For example, in theory, it will eventually be possible to check the genes in a human zygote, to make sure that there are no major genetic faults. If a fault was found, then the 'right' sort of gene could be inserted into the zygote, so that all the cells in the embryo which developed from it contained this 'right' gene. Should this be allowed? Should it be allowed just for very unpleasant diseases such as cystic fibrosis? Or should it be allowed for things like hair colour or anything else? This is all a long way in the future, but perhaps we should think about it now, before things begin to happen that we are not prepared for.

Like all major new scientific discoveries and inventions, genetic engineering has tremendous potential to provide all sorts of benefits for people, and perhaps also for other living things, as well as equally tremendous potential for harm. Scientists and everyone else must remain well aware of this. If as many people as possible, whether they are scientists or not, try to stay well informed about what developments are taking place, then we can do our best to ensure that the 'good' uses of genetic engineering go ahead, while the potential problems are stopped in their tracks.

Chapter revision questions

1. (a) What is a microorganism?
   (b) Describe how one named bacterium is used to make food eaten by humans.

2. Describe the roles of microorganisms in the recycling of water from sewage.

3. Outline the roles of yeast in the production of:
   (a) alcohol
   (b) bread.

4. (a) Describe how mycoprotein is made.
   (b) Why is mycoprotein a useful component of a balanced diet?

5. Describe how
   (a) yoghurt
   (b) cheese
   are produced.

6. (a) Describe how insulin has been produced, using genetic engineering.
   (b) What are the advantages of producing insulin in this way?

15 Living organisms in their environment

15.1 Organisms interact with their environment.

One very important way of studying living things is to study them where they live. Animals and plants do not live in complete isolation. They are affected by their surroundings, or **environment**. Their environment is also affected by them. The study of the interaction between living organisms and their environment is called **ecology**.

15.2 Ecology uses special terms.

There are many words used in ecology with which you need to be familiar. The area where an organism lives is called its **habitat**. The habitat of a tadpole might be a pond. There will probably be many tadpoles in the pond, forming a **population** of tadpoles. A population is a group of organisms of the same species.

But tadpoles will not be the only organisms living in the pond. There will be many other kinds of animals and plants making up the pond **community**. A community is all the organisms, of all the different species, living in the same habitat.

The living organisms in the pond, the water in it, the stones and the mud at the bottom, make up an **ecosystem**. An ecosystem consists of a community and its environment (Fig 15.1).

Within the ecosystem, each living organism has its own life to live and role to play. The way in which an organism lives its life in an ecosystem is called its **niche**. Tadpoles, for example, eat algae and other weeds in the pond; they disturb pebbles and mud at the bottom of shallow areas in the pond; they excrete ammonia into the water; they breathe in oxygen from the water, and breathe out carbon dioxide. All these things, and many others, help to describe the tadpoles' role, or niche, in the ecosystem.

> ### Questions
> 1 What is ecology?
> 2 What is a population?
> 3 Give two examples of an ecosystem, other than a pond.

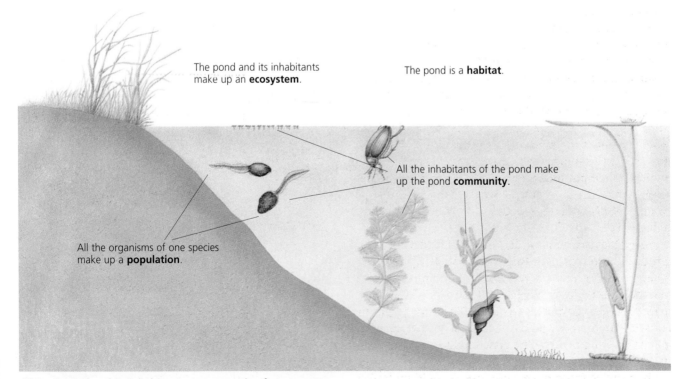

15.1 A pond and its inhabitants – an example of an ecosystem.

Studying ecosystems

15.3 Keys are used for identification.

You may be able to study an ecosystem near to your school.

There are many different ways of studying ecosystems. But whatever ecosystem you study, and however you decide to study it, you will have to begin by identifying the living organisms in it.

Some of them, particularly the bigger ones, you may be able to identify quite quickly from pictures in books. But there will almost certainly be many that you cannot find pictures of, or for which you are not sure if the picture really does show your plant or animal.

When this happens, you will need to use a **key**. A key is a way of leading you through to the name of your organism by giving you two descriptions at a time, and asking you to choose between them. Each choice you make then leads you onto another pair of descriptions, until you end up with the name of your organism.

An example of a key like this is shown in Fig 15.2. It is called a **dichotomous key**, because each time you choose between *two* descriptions ('di' means two).

Food and energy in an ecosystem

15.4 Energy passes along food chains.

All living organisms need energy. They get energy from food, by respiration. All the energy in an ecosystem comes from the sun. Some of the energy in sunlight is captured by plants, and used to make food – glucose, starch and other organic substances such as fats and proteins. These contain some of the energy from the sunlight. When the plant needs energy, it breaks down some of this food by respiration.

Animals get their food, and therefore their energy, by eating plants, or by eating animals which have eaten plants.

The sequence by which energy, in the form of food, passes from a plant to an animal and then to other animals, is called a **food chain**. Fig 15.3 shows one example of a food chain.

15.5 Consumers use food made by producers.

Every food chain begins with green plants because only they can capture the energy from sunlight. They are called **producers**, because they produce food.

Animals are **consumers**. An animal which eats plants is a **primary consumer**, because it is the first consumer in a food chain. An animal which eats that animal is a **secondary consumer**, and so on along the chain.

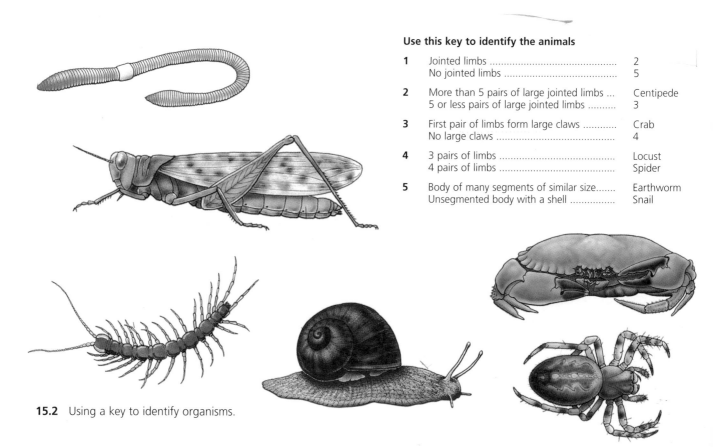

Use this key to identify the animals

1	Jointed limbs	2
	No jointed limbs	5
2	More than 5 pairs of large jointed limbs ...	Centipede
	5 or less pairs of large jointed limbs	3
3	First pair of limbs form large claws	Crab
	No large claws	4
4	3 pairs of limbs	Locust
	4 pairs of limbs	Spider
5	Body of many segments of similar size.......	Earthworm
	Unsegmented body with a shell	Snail

15.2 Using a key to identify organisms.

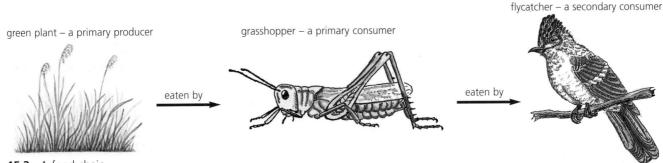

15.3 A food chain.

15.6 Food chains are usually short.

As the energy is passed along the chain, each organism uses some of it. So the further along the chain you go, the less energy there is. There is plenty of energy available for producers, so there are usually a lot of them. There is less energy for primary consumers, and less still for secondary consumers. This means that towards the end of the food chain, the organisms get fewer in number, or smaller in total size (Fig 15.4).

The loss of energy along the food chain also limits the length of it. There are rarely more than five links in a chain, because there is not enough energy left to supply the next link. Many food chains only have three links.

15.7 Consumers feed at different trophic levels.

In Fig 15.5, the number of organisms in the food chain is shown as a pyramid. The size of each block in the pyramid represents the number of organisms. It is called a **pyramid of numbers**. Each level in the pyramid is called a **trophic level** ('trophic' means feeding).

The pyramid is this shape because there is less energy available as you go up the trophic levels, so there are fewer organisms at each level.

Many organisms feed at more than one trophic level. You, for example, are a primary consumer when you eat vegetables, a secondary consumer when you eat meat or drink milk, and a tertiary consumer when you eat a predatory fish such as a salmon.

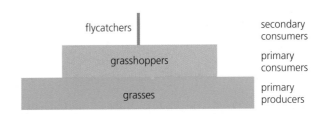

Each level in the pyramid is called a trophic level. The size of each level represents the numbers of organisms feeding at that level.

15.5 A pyramid of numbers.

15.8 Pyramids of numbers may be 'upside-down'.

Fig 15.6 shows a different shaped pyramid of numbers. The pyramid is this shape because of the sizes of the organisms in the food chain. Although there is only a single tree, it is huge compared with the caterpillars which feed on it.

If you make the size of the blocks represent the **mass** of the organisms, instead of their **numbers**, then the pyramid becomes the right shape again. It is called

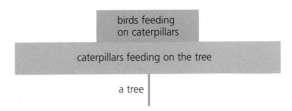

The pyramid is this shape because one tree may provide food for hundreds of caterpillars.

15.6 An inverted pyramid of numbers.

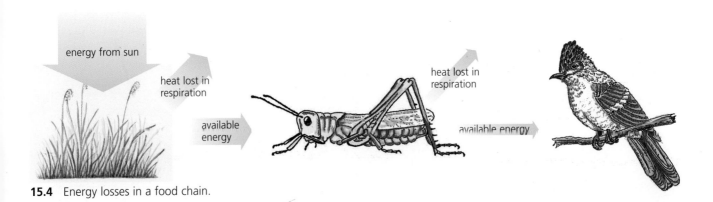

15.4 Energy losses in a food chain.

a **pyramid of biomass** (Fig 15.7), and gives a much better idea of the actual quantity of animal or plant material at each trophic level.

Yet another possibility is to draw a **pyramid of energy**, where the size of the blocks represents the quantity of energy in the organisms.

```
                    birds feeding
                    on caterpillars
caterpillars feeding on the tree
                    a tree
```

In this pyramid, the size of each box represents the mass of each kind of organism.

15.7 A pyramid of biomass.

15.9 Understanding energy flow helps agriculture.

Understanding how energy is passed along a food chain can be useful in agriculture. We can eat a wide variety of foods, and can feed at several different trophic levels. Which is the most efficient sort of food for a farmer to grow, and for us to eat?

The nearer to the beginning of the food chain we feed, the more energy there is available for us. This is why our staple foods such as, wheat, rice and potatoes, are plants.

When we eat meat, eggs or cheese or drink milk, we are feeding further along the food chain. There is less energy available for us from the original energy provided by the sun. It would be more efficient in principle to eat the grass in a field, rather than to let cattle eat it, and then eat them.

In fact, however, although there is far more energy in the grass than in the cattle, it is not available to us. We simply cannot digest the cellulose in grass, so we cannot release the energy from it. The cattle can; they turn the energy in cellulose into energy in protein and fat, which we can digest.

However, there are many plant products which we *can* eat. Soya beans, for example, yield a large amount of protein, much more efficiently and cheaply than cattle or other animals. A change towards vegetarianism would enable more food to be produced on the Earth, if the right crops were chosen.

Food chains link to form food webs. In the food web shown below in Fig 15.8, there is still too little energy available to support more than a few animals at the ends of longer food chains, such as a bird that feeds on fish and other aquatic animals.

> **Questions**
> 1 Where does all the energy in living organisms originate from?
> 2 Write down a food chain (a) which ends with humans, (b) in the sea and (c) with five links in it.
> 3 Why are green plants called producers?
> 4 Why are there rarely more than five links in a food chain?

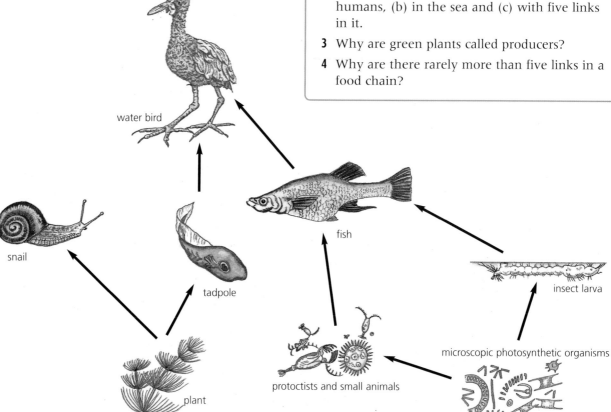

15.8 A freshwater food web.

Nutrient cycles

15.10 Decomposers release minerals from dead organisms.

One very important group of organisms which it is easy to overlook when you are studying an ecosystem, is the **decomposers**. They feed on waste material from animals and plants, and on their dead bodies. Many fungi and bacteria are decomposers.

Decomposers are extremely important, because they help to release substances from dead organisms, so that they can be used again by living ones. Two of these substances are carbon and nitrogen.

15.11 Carbon is recycled.

Carbon is a very important component of living things, because it is an essential part of carbohydrates, fats and proteins.

Fig 15.9 shows how carbon circulates through an ecosystem. The air contains about 0.04 % carbon dioxide. When plants photosynthesise, carbon atoms from carbon dioxide become part of glucose or starch molecules in the plant.

Some of the glucose is then broken down by the plant in respiration. The carbon in the glucose becomes part of a carbon dioxide molecule again, and is released back into the air.

Some of the carbon in the plant will be eaten by animals. The animals respire, releasing some of it back into the air as carbon dioxide.

When the plant or animal dies, decomposers will feed on them. The carbon becomes part of the decomposers' bodies. When they respire, they release carbon dioxide into the air again.

15.12 Few organisms can use nitrogen gas.

Living things need nitrogen to make proteins. There is plenty of nitrogen around. The air is about 79 % nitrogen gas. Molecules of nitrogen gas, N_2, are made of two nitrogen atoms joined together. These molecules are very inert, which means that they will not readily react with other substances.

So, although the air is full of nitrogen, it is in such an unreactive form that plants and animals cannot use it at all. It must first be changed into a more reactive form, such as ammonia (NH_3) or nitrates (NO_3^-).

Changing nitrogen gas into a more reactive form is called **nitrogen fixation**. There are several ways that it can happen.

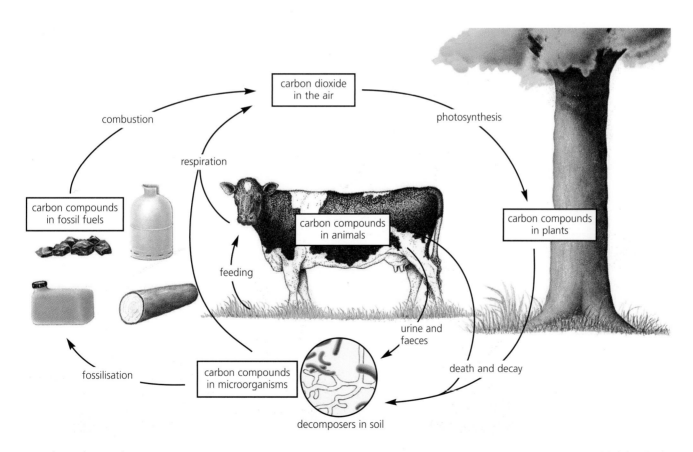

15.9 The carbon cycle.

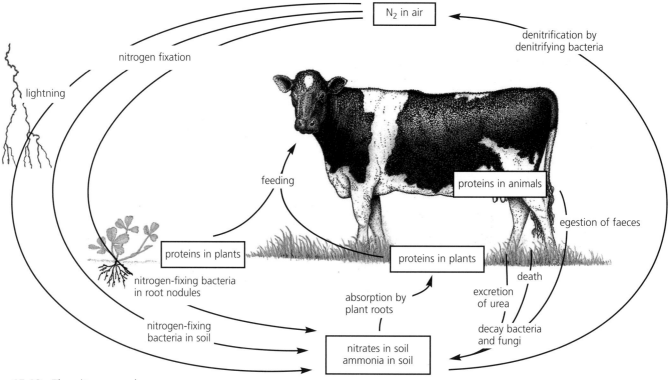

15.10 The nitrogen cycle.

Lightning

Lightning makes some of the nitrogen gas in the air combine with oxygen, forming nitrogen oxides. They dissolve in rain, and are washed into the soil, where they form nitrates.

Artificial fertilisers

Nitrogen and hydrogen can be made to react in an industrial chemical process, forming ammonia. The ammonia is used to make ammonium compounds and nitrates, which are sold as fertilisers.

Nitrogen-fixing bacteria

These bacteria live in the soil, or in root nodules (small swellings) on plants like peas, beans and clover. One kind is called *Rhizobium* ('rhizo' means root, 'bium' means living). They use nitrogen gas from the air spaces in the soil, and combine it with other substances to make nitrates and other compounds.

15.13 Fixed nitrogen moves round the nitrogen cycle.

Once the nitrogen has been fixed, it can be absorbed by the roots of plants, and used to make proteins. Animals eat the plants, so animals get their nitrogen in the form of proteins.

When an animal or plant dies, bacteria and fungi decompose the body. The protein, containing nitrogen, is broken down to ammonia and this is released. Another group of bacteria, called **nitrifying bacteria**, turn the ammonia into nitrates, which plants can use again.

Nitrogen is also returned to the soil when animals excrete nitrogenous waste material. It may be in the form of ammonia or urea. Again, nitrifying bacteria will convert it to nitrates.

15.14 Denitrifying bacteria make nitrogen gas.

A third group of bacteria complete the nitrogen cycle. They are called **denitrifying bacteria**, because they undo the work done by nitrifying bacteria. They turn nitrates and ammonia in the soil into nitrogen gas, which goes into the atmosphere (Fig 15.10).

15.15 Carnivorous plants get nitrogen from insects.

If the soil is waterlogged, nitrogen-fixing bacteria cannot live there, but denitrifying ones can. So boggy soil is usually very short of nitrates. Plants living in these places either have to manage with very little nitrogen, or get it from somewhere else. Some of them have become carnivorous. Plants like the Venus fly trap, or the sundews, supplement their diet with insects. They digest them with enzymes, and get extra nitrogen from the protein in the insects' bodies.

> **Questions**
> 1 What is a decomposer?
> 2 Why are decomposers important?

15.16 The water cycle.

Fig 15.11 shows how water cycles between living organisms and their environment. Living things, especially trees, play a very important role in this cycle. When rain falls, the tree roots absorb water from the soil. The water travels up through xylem vessels into the leaves, where some of it evaporates and diffuses out of the stomata as water vapour.

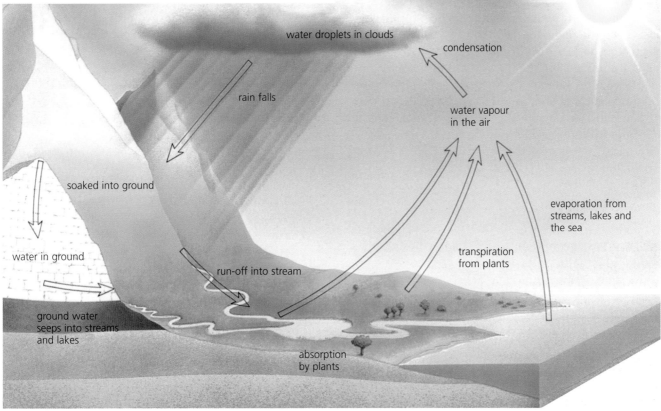

15.11 The water cycle.

Population size

15.17 Most populations stay about the same size.

We have seen that a population is all the individuals of a particular species that live together in a habitat. In this section, we will look at how and why population sizes change, and begin to consider the implication of Earth's rapidly increasing human population.

Most populations tend to stay roughly the same size over a period of time. They may go up and down, or fluctuate, but the average population will probably stay the same over a number of years. The population of greenfly in a garden, for example, might be much greater one year than the next. But their numbers will almost certainly be back to normal in a year or so. Over many years, the sizes of most populations tend to remain at around the same level.

Yet if all the offspring of one female greenfly survived and reproduced, she could be the ancestor of 600 000 000 000 greenfly in just one year! Why doesn't the greenfly population shoot upwards like this? Why isn't the world overrun with greenfly?

The answers to those questions are of great importance to human beings, because our own population is doing just that; it is shooting upwards at an alarming rate. Every hour, there are 9000 extra people in the world. We need to understand why this is happening, and what is likely to happen next. Can we slow down the increase? What happens if we don't?

15.18 Birth rate and death rate determine population size.

The size of a population depends on how many individuals leave the population, and how many enter it.

Individuals leave a population when they die, or when they migrate to another population. Individuals enter a population when they are born, or when they migrate into the population from elsewhere. Usually, births and deaths are more important in determining population sizes than immigration and emigration.

A population increases if new individuals are born faster than the old ones die, that is when the birth rate is greater than the death rate. If birth rate is less than death rate, then the population will decrease. If birth rate and death rate are equal, the population will stay the same size.

This explains why we are not knee-deep in greenfly. Although the greenfly population's birth rate is enormous, the death rate is also enormous. Greenfly are eaten by ladybirds and birds, and sprayed by gardeners. Over a period of time, the greenfly's birth and death rates stay about the same, so the population doesn't change very much.

15.19 Yeast experiments give some clues about population growth.

By looking at changes in population sizes in other organisms, we can learn quite a lot about our own. Many experiments on population sizes have been done on organisms like bacteria and yeast, because they reproduce quickly and are easy to grow. Fig 15.12 shows the results of an experiment in which a few yeast cells are put into a container of nutrient broth. The cells feed on the broth, grow and reproduce. The numbers of yeast cells are counted every few hours.

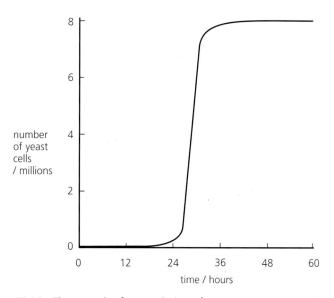

15.14 The growth of a population of yeast.

At the beginning of the experiment, the population only grows quite slowly, because there are not many cells there to reproduce. But once they get going, growth is very rapid. Each cell divides to form 2, then 4, then 8, then 16. There is nothing to hold them back except the time it takes to grow and divide.

But as the population gets larger, the individual cells can no longer reproduce as fast, and begin to die off more rapidly. This may be because there is not enough food left for them all, or it might be that they have made so much alcohol that they are poisoning themselves. The cells are now dying off as fast as new ones are being produced, so the population stops growing and levels off.

15.20 Environmental factors control population size.

Although the experiment with the yeast is done in artificial conditions, a similar pattern is found in the growth of populations of many species in the wild. If a few individuals get into a new environment, then their population curve may be very like the one for yeast cells in broth. The population increases quickly at first, and then levels off.

The levelling off is always caused by some kind of environmental factor. In the case of the yeast, the factor may be food supply. Other populations may be limited by disease, or the number of nest sites, or the number of predators, for example. The factor that stops the population from getting any larger is called a **limiting factor**.

15.21 Population sizes often oscillate.

It is usually very difficult to find out which environmental factors are controlling the size of a population. Almost always, many different factors will interact. A population of rabbits, for example, might be affected by the number of foxes and other predators, the amount of food available, the amount of space for burrows, and the amount of infection by the virus which causes myxomatosis.

Fig 15.13 shows an example of how the size of population of a predator may be affected by its prey. This information comes from the number of skins which were sold by fur traders in Northern Canada to the Hudson Bay Company, between 1845 and 1925. Snowshoe hares and northern lynxes were both trapped for their fur, and the numbers caught probably give a very good idea of their population sizes.

Snowshoe hare populations tend to vary from year to year. No-one is quite sure why this happens, but it may be related to their food supply. Whenever the snowshoe hare population rises, the lynx population also rises shortly afterwards, as the lynxes now have more food. A drop in the snowshoe hare population is rapidly followed by a drop in the lynx population.

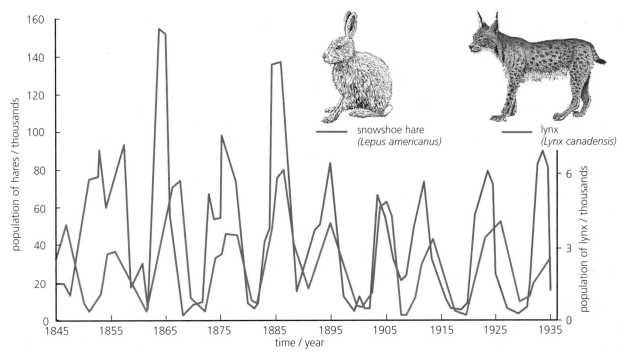

15.13 Variations in snowshoe hare and lynx populations in Northern Canada.

The numbers tend to go up and down, or **oscillate**, but the average population sizes stay roughly the same over many years.

15.22 Age pyramids show whether a population is increasing or decreasing.

When scientists begin to study a population, they want to know whether the population is growing or shrinking. This can be done by counting the population over many years, or by measuring its birth rate and death rate. But often it is much easier just to count the numbers of individuals in various age groups, and to draw an **age pyramid**.

Fig 15.14 shows two examples of age pyramids. The size of each box represents the numbers of individuals of that age.

Fig 15.14a is a bottom-heavy pyramid, because there are far more young individuals than old ones. This indicates that birth rate is greater than death rate, so this population is increasing.

Fig 15.14b shows a much more even spread of ages. Birth rate and death rate are probably about the same. This population will remain about the same size.

If an age pyramid is drawn for the human population on Earth, it is bottom-heavy, like Fig 15.14a. Age pyramids for most of the world's developing countries are also this shape, showing that their populations are increasing. But an age pyramid for a European country such as France looks more like Fig 15.14b. The human population in France is staying about the same.

a An increasing population
If all the organisms in the younger age groups grow up and reproduce, the population will increase.

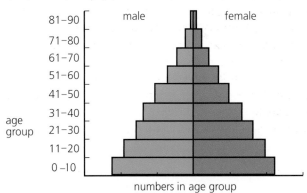

b A stable population
The sizes of the younger age groups are only a little larger than the older ones, so this population should not change much in size.

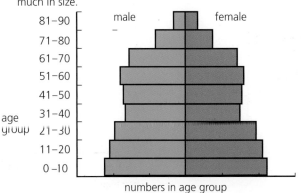

15.14 Age pyramids.

167

15.23 The human death rate has decreased dramatically.

Fig 15.15 shows how the human population of the world has changed since about 3000 BC. For most of that time, human populations have been kept in check by a combination of disease, famine and war. Nevertheless, there has still been a steady increase.

Twice there have been definite 'spurts' in this growth. The first was around 8000 BC, not shown on the graph, when people in the Middle East began to farm, instead of just hunting and finding food. The second began around 300 years ago, and is still happening now.

There are two main reasons for this recent growth spurt. The first is the **reduction of disease**. Improvements in water supply, sewage treatment, hygienic food handling and general standards of cleanliness have virtually wiped out many diseases in countries such as the USA and most European countries – for example typhoid and dysentery. Immunisation against diseases such as polio has made these very rare indeed. Smallpox has been totally eradicated. And the discovery of antibiotics has now made it possible to treat most diseases caused by bacteria.

Secondly, there has been an **increase in food supply**. More and more land has been brought under cultivation. Moreover, agriculture has become more efficient, so that in many parts of the world each hectare of land is now producing more than ever before.

15.24 Birth rate now exceeds death rate.

The human population has increased dramatically because the death rate has been brought down. More and more people are now living long enough to reproduce. If the birth rate doesn't drop by the same amount as the death rate, then the world population will continue to increase.

In developed countries, the dramatic fall in the death rate began in about 1700. To begin with, the birth rate stayed high, so the population grew rapidly. But since 1800, there has been a marked drop in birth rate. In 1870, for example, the 'average' British family was 6.6 children, but by 1977 it was only 1.8. In Britain, birth rate and death rate are now about equal, so the population is staying the same.

However, in many of the developing countries, the fall in the death rate only began about 50 years ago. As yet, the birth rates have not dropped, and so the populations are rising rapidly.

15.25 Birth rate must be reduced to slow population growth.

The human population could be brought back under control in two ways – increasing the death rate or decreasing the birth rate. There is no question as to which of these is the best.

In the developed countries, the single largest factor which brought down the birth rate was the introduction of contraceptive techniques. Considerable efforts are being made to introduce these to people in the developing countries, with some success. But many people are suspicious of contraceptive methods, or barred from using them by their religion, or simply want to have large families. It looks as though the population will go on rising for at least another 200 years.

If we do not control the overall human birth rate, then it may happen that famine, war or disease will increase the death rate. This cannot be the best thing for the human race. We must do our best to stabilise the world population at a level at which everyone has a fair chance of a long, healthy life.

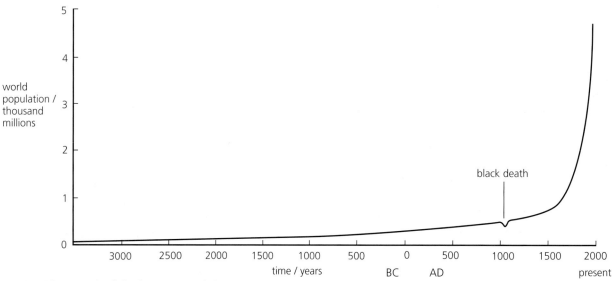

15.15 The growth of the human population.

Chapter revision questions

1. (a) Why do living organisms need carbon?
   (b) Explain how carbon atoms become part of a plant.
   (c) What happens to some of these carbon atoms when a plant respires?
   (d) Explain the role of decomposers in the carbon cycle.

2. (a) Why do living organisms need nitrogen?
   (b) Explain why plants and animals cannot use the nitrogen in the air
   (c) What is nitrogen fixation?
   (d) Where do nitrogen-fixing bacteria live?
   (e) Explain how do animals obtain nitrogen
   (f) What do nitrifying bacteria do?
   (g) Which type of bacteria return nitrogen to the air?

3. Explain the difference between each of the following pairs, giving examples where you can.
   (a) producer, consumer (b) primary consumer, secondary consumer (c) community, population (d) food chain, food web (e) pyramid of biomass, pyramid of numbers

4. Construct a dichotomous key, to enable someone to identify six people in your class.

5. Using the following list of words, in order, explain how (a) a carbon atom in the air becomes part of a glucose molecule in your biceps muscle, and (b) how that carbon atom might return to the air again.

   broad bean plant, stomata, photosynthesis, glucose, sucrose, phloem vessel, bean seed, starch, feeding, amylase, maltose, maltase, glucose, ileum, hepatic portal vein, liver, hepatic vein, heart, aorta, subclavian artery, capillary, diffusion, muscle, respiration, carbon dioxide, diffusion, capillary, subclavian vein, heart, pulmonary artery, capillary, diffusion, alveolus, expiration

6. (a) Why is nitrogen important to living organisms?
   (b) In what form do each of the following obtain their nitrogen? (i) a green plant
   (ii) nitrogen-fixing bacteria (iii) a mammal
   (c) In the sea, the main nitrogen-fixing organisms are blue-green algae, which float near the top of the water in the plankton. Construct a diagram or chart similar to Fig 15.10, showing how nitrogen is circulated amongst marine organisms.

7. A fish tank was filled with water, and some bacteria were added. Some phytoplankton (microscopic plants) were then introduced. The tank was put into a dark place, and left for eight months.

   At intervals, the water was tested to find out what it contained. The results are shown in the graph in Fig 15.16.

   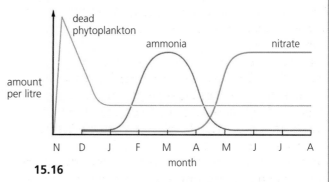
   15.16

   (a) Why did the phytoplankton die so quickly?
   (b) The phytoplankton contain nitrogen in their cells. In what form is most of this nitrogen?
   (c) Why does the quantity of dead phytoplankton decrease during the first two months of the experiment?

   After one month, ammonia begins to appear in the water. (d) Where has this ammonia come from? (e) What kind of bacteria are responsible for its production? (f) When does nitrate begin to appear in the water? (g) What kind of bacteria are responsible for its production?

8. The graph in Fig 15.17 shows population changes over one summer, for two insects. One is a type of greenfly, and the other is a ladybird which feeds on it.

   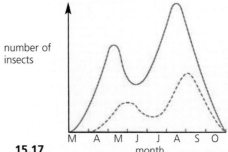
   15.17

   (a) Which curve represents the ladybird population, and which the greenfly population?
   (b) Give a reason for your answer to part (a).
   (c) Explain why the two curves are similar.
   (d) Why do the two curves rise and fall at slightly different times?

16 Humans and the environment

16.1 All organisms affect their environment.

All living things affect the living and non-living things around them. For example, earthworms make burrows and wormcasts, which affect the soil, and therefore the plants growing in it. Rabbit fleas carry the virus which causes myxomatosis, so they can affect the size of a rabbit population, and perhaps the size of the fox population if the foxes depend on rabbits for food.

Perhaps the biggest ever effect of living organisms on the environment happened about 1500 million years ago. At this time, the first living cells which could photosynthesise evolved. Until then, there had been no oxygen in the atmosphere. These organisms began to produce oxygen, which gradually accumulated in the atmosphere. The oxygen in the air we now breathe has been produced by photosynthesis. The appearance of oxygen in the air meant that many anaerobic organisms could now only live in particular parts of the Earth which were oxygen-free, such as in deep layers of mud. It meant that many other kinds of organism could evolve, which used the oxygen for respiration. All this oxygen excreted by photosynthetic organisms could be considered to be the biggest case of pollution ever!

16.2 Humans affect the environment.

Within the past 10 000 years or so, another organism has had an enormous impact on the environment. Ever since humans learnt to hunt with weapons, to domesticate animals and to farm crops, we have been changing the environment around us in a very significant way. We kill wild animals for food, decreasing their populations and making some species extinct. We cut down forests. We build cities, roads and dams. We release harmful substances into the water, air and soil.

As the human population continues to increase, and expectations of living standards become higher, the effects which we have on the environment become ever greater. It is very important that we do our best to understand what we are doing, and what the effects might be. The more we understand, the more we can do to prevent too much damage before it happens, and keep the Earth a pleasant place for humans, plants and other animals to live.

Table 16.1 summarises some of the damaging effects we have had on our environment. The rest of this chapter describes some of these in more detail, and explains what we can do to limit further damage.

Table 16.1 A summary of the harmful effects of humans on the environment.

Damage	Example	Main causes	Possible solutions
Air pollution	Damage to the ozone layer	CFCs	Stop using CFCs; find harmless alternatives
	Global warming	Enhanced greenhouse effect, caused by release of carbon dioxide, methane, CFCs and nitrogen oxides	Reduce use of fossil fuels; stop using CFCs; produce less organic waste and/or collect and use methane produced from landfill sites
	Acid rain	Sulphur dioxide and nitrogen oxides from the burning of fossil fuels	Burn less fossil fuel. Use catalytic converters on cars
Habitat destruction	Deforestation	Destruction of forests, especially rainforests, for wood and for land for farming, roads and houses	Provide alternative sources of income for people living near rainforests
	Loss of wetlands	Draining wetlands for housing and land for farming	Protect areas of wetland
Water pollution	Toxic chemicals	Untreated effluent from industry; run-off from mining operations	Impose tighter controls on industry and mining
	Eutrophication	Sewage and fertilisers running into streams	Treat all sewage before discharge into streams; use fewer fertilisers
	Oil spills	Shipwrecks; leakages from undersea oil wells	Impose tighter controls on shipping and the oil industry
Species destruction	Loss of habitat	See deforestation and wetlands above	See above
	Damage from pesticides	Careless use of insecticides and herbicides	Development of more specific and less persistent pesticides; more use of alternative control methods, such as biological control
	Damage from fishing	Overfishing, greatly reducing populations of species caught for food; accidental damage to other animals, such as dolphins	Impose controls on methods and amount of fishing

Global warming

16.3 The greenhouse effect is essential to life.

The Earth's atmosphere contains several different gases which act like a blanket, keeping the Earth warm. The most important of these gases is **carbon dioxide**.

Carbon dioxide is transparent to shortwave radiation from the Sun. The sunlight passes freely through the atmosphere (Fig 16.1), and reaches the ground. The ground is warmed by the radiation, and emits longer wavelength, infrared radiation. Carbon dioxide does not let all of this infrared radiation pass through. Much of it is kept in the atmosphere, making the atmosphere warmer.

This is called the **greenhouse effect**, because it is just the same as the effect which keeps an unheated greenhouse warmer than the air outside. The glass around the greenhouse behaves like the carbon dioxide in the atmosphere. It lets shortwave radiation in, but does not let out the longwave radiation. The longwave radiation is trapped inside the greenhouse, making the air inside it warmer.

We need the greenhouse effect. If it did not happen, then the Earth would be frozen and lifeless. The average temperature on Earth would be about 33 °C lower than it is now.

16.4 The enhanced greenhouse effect may cause global warming.

Although we need the greenhouse effect, people are worried that it may be increasing. The amount of carbon dioxide and other greenhouse gases in the atmosphere is getting greater. This may trap more infrared radiation, and make the atmosphere warmer. This is called the **enhanced greenhouse effect**, and its possible effect on the Earth's temperature is called **global warming**.

Over recent years, the amount of fossil fuels which have been burnt by industry, and in engines of vehicles such as cars, trains and aeroplanes, has increased greatly. This releases carbon dioxide into the atmosphere. Fig 16.2 (overleaf) shows what has happened to the amount of carbon dioxide in the atmosphere since 1750.

Other gases which contribute to the greenhouse effect have also been released by human activities. These include **methane**, **nitrogen oxides** and **CFCs**. Table 16.2 shows where these gases come from. The concentrations of all of these gases in the atmosphere is steadily increasing.

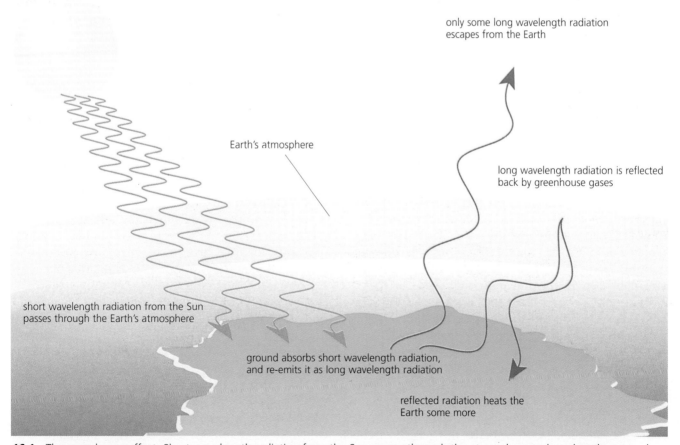

16.1 The greenhouse effect. Short wavelength radiation from the Sun passes through the atmosphere and reaches the ground. Some of it is absorbed by the ground, and is re-emitted as longwave radiation. Much of this cannot pass through the blanket of 'greenhouse gases' in the atmosphere. It is reflected back towards the Earth, warming the atmosphere.

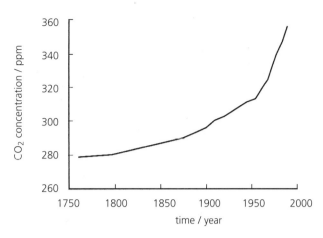

16.2 How carbon dioxide concentrations in the atmosphere have changed since 1750.

Table 16.2 Gases contributing to the greenhouse effect.

Gas	% Estimated contribution	Main sources
Carbon dioxide	55	Burning fossil fuels
Methane	15	Decay of organic matter, e.g. in waste tips and paddy fields; waste gases from digestive processes in cattle and insects; natural gas leaks
CFCs	24	Fridges and air conditioning systems; plastic foams
Nitrogen oxides	6	Fertilisers; burning fossil fuels, especially in vehicles

Some scientists think that, as the concentration of these gases increases, the temperature on Earth will also increase. At the moment, however, no-one is quite sure whether this will happen or not. There are all sorts of other processes, many of them natural, which can cause quite large changes in the average temperature of the Earth, and these are not fully understood. For example, every now and then the Earth has been plunged into an Ice Age. Perhaps we are due for another Ice Age soon. Perhaps the enhanced greenhouse effect might help to delay this!

But most people think that we should be very worried about the enhanced greenhouse effect and global warming. If the Earth's temperature did rise significantly, there could be big changes in the world as we know it. For example, the ice caps might melt. This would release a lot more water into the oceans, so that sea levels would rise. Many low-lying areas of land might be flooded. This could include large parts of countries like Bangladesh, almost the whole of the Maldive islands, and major cities such as London.

A rise in temperature would also affect the climate in many parts of the world. No-one is sure just what would happen where – there are too many variables for scientists to be able to predict the consequences. It would probably mean that some countries which already have low rainfall might become very dry deserts. Others might have more violent storms than they do now. This would mean that animals and plants living in some areas of the world might become extinct. People in some places might not be able to grow crops.

There might be some beneficial effects, too. For example, extra carbon dioxide in the atmosphere and higher temperatures might increase the rate of photosynthesis in some parts of the world. This could mean that higher yields could be gained from crops.

16.5 Can we avoid global warming?

The fact is that, although we know for sure that the levels of greenhouse gases are increasing, we do *not* know for sure whether this will cause global warming. Nor do we know for sure what effects global warming would have. But, for the very reason that we do not know these things, we need to take great care with our atmosphere. We cannot afford to do this enormous experiment with our planet!

It is important that we cut down the emission of greenhouse gases. One obvious way to do this is to reduce the amount of **fossil fuels** which are burnt. This would reduce the amount of **carbon dioxide** we pour into the air. Agreements have been made between countries to try to do this, but they are proving very difficult to implement.

Deforestation has also been blamed for increasing the amount of carbon dioxide in the air. (You can read more about deforestation in Sections 16.10–16.12.) It has been argued that, by cutting down rainforests, there are fewer trees to photosynthesise and remove carbon dioxide from the air. However, this is not straightforward. A mature forest tree gives out almost as much carbon dioxide from respiration as it takes in by photosynthesis! Also, when the tree has been cut down, other plants grow in its place – either naturally, or planted for crops by farmers. These probably take as much carbon dioxide from the air as the tree did.

However, if the tree is burnt or left to rot when it is chopped down, then carbon dioxide will be released from it. So just chopping down trees does not increase the amount of carbon dioxide in the air, but burning them or letting them decay does (Fig 16.6).

Farming can add **nitrogen oxides** to the air. This is because nitrogen-containing fertilisers can release this gas. Nitrogen oxides also come from burning fossil fuels. We could reduce the amount of nitrous oxide which we release by reducing the usage of fertilisers, as well as cutting down the use of fossil fuels.

Methane, like nitrogen oxides, is produced by farming activities. It is released by bacteria which live on organic matter, such as in paddy fields (flooded fields which are used for growing rice), by animals which chew the cud, such as cattle, and by some insects, such as termites. There is probably not much that we can do about this. Methane is also produced by decaying rubbish in landfill sites. We can help this problem by reducing the amount of rubbish which we throw away, and by collecting the methane from these sites. It can be used as fuel (Fig 16.4). Although burning it for fuel does release carbon dioxide, this carbon dioxide does not trap so much infrared radiation as the methane would have done.

16.4 Methane being collected from a landfill site. The methane is used to fuel an electricity generator.

16.3 Methane produced by microorganisms living in paddy fields, such as these in Thailand.

Questions

1 Explain the difference between **the greenhouse effect**, **the enhanced greenhouse effect** and **global warming**.

2 Each of the following has been suggested as a way of reducing global warming. For each suggestion, explain why it would work, and discuss the problems which would probably occur in trying to implement it.
(a) reducing the top speed limit for cars and trucks
(b) improving traffic flow in urban areas
(c) insulating houses in countries with cold climates
(d) increasing the number of nuclear power stations
(e) encouraging people to recycle more of their rubbish

Acid rain

16.6 Burning fossil fuels releases sulphur and nitrogen oxides.

Fossil fuels, such as coal, oil and natural gas, were formed from living organisms. They all contain sulphur; coal contains the most. When they are burnt, the sulphur combines with oxygen in the air and forms sulphur dioxide. Nitrogen oxides are also formed.

Sulphur dioxide is a very unpleasant gas. If people breathe it in, it can irritate the linings of the breathing system. If you are prone to asthma or bronchitis, sulphur dioxide can make it worse. Sulphur dioxide is also poisonous to many kinds of plants, sometimes damaging their leaves so badly that the whole plant dies.

16.7 Sulphur and nitrogen oxides produce acid rain.

Rain is usually slightly acid, with a pH a little below 7. This is because carbon dioxide dissolves in it to form carbonic acid.

Sulphur dioxide and nitrogen oxides also dissolve in rain. They form an acidic solution, called **acid rain**. The pH of acid rain can be as low as 4.

Acid rain damages plants. Although the rain usually does not hurt the leaves directly when it falls onto them, it does affect the way in which plants grow. This is because it affects the soil in which the plants are growing. The acid rain water seeps into the soil, and washes out ions such as calcium, magnesium and aluminium. The soil becomes short of these ions, so the plant becomes short of nutrients. It also makes it more difficult for the plant to absorb other nutrients from the soil. So acid rain can kill trees and other plants.

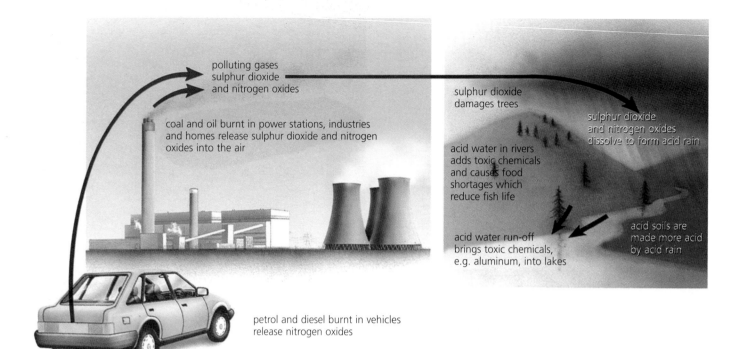

16.5 Acid rain.

The ions which are washed out of the soil by the acid rain often end up in rivers and lakes. Aluminium, in particular, is very poisonous to fish, because it affects their gills. Young fish are often killed if the amount of aluminium in the water is too great. Other freshwater organisms are often killed, too. At the same time, the water itself becomes more acidic, which means that many kinds of plants and animals cannot live in it (Fig 16.5).

16.8 Sulphur dioxide is carried long distances.

One of the biggest problems in trying to do anything about the problems of acid rain is that it does not usually fall anywhere near the place which is causing it. A coal-burning power station might release a lot of sulphur dioxide, which is then carried high in the air for hundreds of miles before falling as acid rain. Sulphur dioxide produced in England might fall as acid rain in Norway.

16.9 Sulphur dioxide emissions are being reduced.

Acid rain is, in many ways, a much easier problem to solve than the the greenhouse effect. The answer is simple – we must cut down emissions of sulphur dioxide and nitrogen oxides.

Coal-burning power stations have been the worst culprits. The number of coal-burning power stations in some European countries has been going down – more because of the cost of coal than to try to stop pollution – and more of them are burning oil which produces less sulphur dioxide. However, this has meant a reduction in the demand for coal, which has been part of the reason for the closure of many coal mines, and the loss of jobs for many miners in countries such as Britain.

Whatever kind of fossil fuel is burnt in power stations or other industries, the waste gases can be 'scrubbed' to remove sulphur dioxide. This often involves passing the gases through a fine spray of lime.

The burning of petrol in car engines also produces nitrogen oxides. These can be removed by catalytic convertors fitted to the exhaust system. All new petrol-burning cars now have to have catalytic converters.

> ### Questions
> 1 What causes acid rain?
> 2 How does acid rain damage trees?
> 3 How does acid rain damage fish?
> 4 Summarise what is being done to try to reduce the production of acid rain.

Nuclear fall-out

16.10 Radiation may cause cancer.

Accidents at nuclear power stations may release radioactive substances into the atmosphere. Exposure to large amounts of radiation from these substances can cause **radiation sickness** and **burns** because cells are badly damaged. This type of radiation can also increase mutation rates, which may lead to cancer.

Deforestation

16.11 People cut down trees for fuel and farmland.

Humans have always cut down trees. Wood is an excellent fuel and building material. The land on which trees grow can be used for growing crops for food, or to sell. One thousand years ago, most of Europe was covered by forests. Now, most of them have been cut down. The cutting down of large numbers of trees is called **deforestation**.

16.7 Unspoiled tropical rainforest in Sarawak.

16.6 When rainforest is cut down and burnt, as here in Brazil, large amounts of carbon dioxide are released and soil nutrients lost.

16.12 Rainforests are special places.

Recently, most concern about deforestation has been about the loss of **tropical rainforests**. In the tropics, the relatively high and constant temperatures, and high rainfall, provide perfect conditions for the growth of plants (Fig 16.7). A rainforest is a very special place, full of many different species of plants and animals. More different species live in a small area of rainforest than in an equivalent area of any other habitat in the world. We say that rainforest has a high **species diversity**.

When an area of rainforest is cut down, the soil under the trees is exposed to the rain. The soil of a rainforest is very thin. The soil is quickly washed away once it loses its cover of plants. This soil erosion may make it very difficult for the rainforest to grow back again, even if the land is left alone. The soil can also be washed into rivers, silting it and causing flooding.

The loss of part of a rainforest means a loss of a habitat for many different species of animals. Even if small 'islands' of forest are left as reserves, these may not be large enough to support a breeding population of the animals. Deforestation threatens many species of animals and plants with extinction.

The loss of so many trees can also affect the water cycle (Fig 15.11). With the trees growing, when rain falls a lot of it is taken up by the trees and taken up to their leaves. It then evaporates, and goes back into the atmosphere in the process of transpiration. If the trees have gone, then the rain simply runs off the soil and into rivers. Much less goes back into the air as water vapour. The air becomes drier, and less rain falls. This can make it much more difficult for people to grow crops and keep livestock.

16.13 Developing countries need help to conserve rainforests.

When people in industrialised countries get concerned about the rate at which some countries are cutting down their forests, it is very important they should remember that they have already cut down most of theirs!

No-one is quite sure how fast the tropical rainforests are being cut down. In the early 1990s, it was thought that tropical rainforest was being cut down at a rate of about 10 or 11 hectares per minute! It is now thought that this was an overestimate – the rate is probably quite a lot less than this. But it is still great enough that we should be very concerned about it.

Most rainforests grow in developing countries, and in some countries many of the people are very poor. The people may cut down the forests to clear land on which they can grow food. It is difficult to expect someone who is desperately trying to produce food, to keep their family alive, not to do this, unless you can offer some alternative. International conservation groups such as the World Wide Fund for Nature, and governments of the richer, developed, countries such as the USA, can help by providing funds to the people or governments of developing countries to try to help them to provide alternative sources of income for people. Many of the most successful projects involve helping local people to make use of the rainforest in a sustainable way.

The greatest pressure on the rainforest may come from the country's government in the big cities, rather than the people living in or near the rainforest. For example, Cameroon in West Africa still has a lot of rainforest. The government of Cameroon desperately needs more money, partly to pay off debts to developed countries. A European country has offered Cameroon large amounts of money in return for being allowed to harvest large areas of rainforest for timber. It is difficult for a country like Cameroon to refuse such an offer. It is important that developed countries should not put these pressures onto them. Instead, the developed countries should try to help the developing countries to conserve their natural resources such as rainforests. In Cameroon, conservation groups are working with local people to help them to live with the rainforest without destroying it. Most local people very much want to do this; they love their forest, and do not want to see it destroyed.

Questions

1 Why are tropical rainforests often considered to be 'very special places'?
2 Explain why people cut down large areas of forest. (You can probably think of other reasons as well as those which are described above.)
3 Outline the damage which can be caused when rainforests are cut down.
4 How can developed countries help to prevent the loss of rainforest?

Water pollution

16.14 Aquatic organisms need clean water.

Many organisms live in water. They are called **aquatic** organisms. Aquatic habitats include fresh water, such as streams, rivers, ponds and lakes; and also marine environments – the sea and oceans.

Most organisms which live in water respire aerobically, and so need oxygen. They obtain their oxygen from oxygen gas which has dissolved in the water. Anything which reduces the amount of oxygen available in the water can make it impossible for fish or other aquatic organisms to live there.

In the UK, there are two main sources of pollution which can reduce oxygen levels in fresh water. They are **fertilisers** and **untreated sewage**.

Farmers and horticulturalists use fertilisers to increase the yield of their crops. The fertilisers usually contain **nitrates** and **phosphates**. Nitrates are very soluble in water. If nitrate fertiliser is put onto soil, it may be washed out in solution when it rains. This is called **leaching**. The leached nitrates may run into streams and rivers.

Algae and green plants in the river grow faster when they are supplied with these extra nitrates. They may grow so much that they completely cover the water. They block out the light for plants growing beneath them, which die. Even the plants on the top of the water eventually die. When they do, their remains are a good source of food for **bacteria**. The bacteria breed rapidly. The large population of bacteria respires, using up oxygen from the water. Soon, there is very little oxygen left for other living things. Those which need a lot of oxygen, such as fish, have to move to other areas, or die.

This whole process is called **eutrophication** (Figs 16.8 and 16.9) It can happen whenever food for plants or bacteria is added to water. As well as fertilisers, other pollutants from farms, such as slurry from buildings where cattle or pigs are kept, or from pits where grass is rotted down to make silage, can cause eutrophication. Untreated sewage can also cause eutrophication (Fig 16.10). Sewage provides a good food source for many kinds of bacteria. Once again, their population grows, depleting the oxygen levels.

16.15 Nitrate use must be carefully controlled.

Eutrophication is not the only problem caused by the leaching of nitrate fertilisers. Some of the nitrates are carried deep into the soil, where they find their way into water in rocks deep underground, called **aquifers**. Water in aquifers may be extracted to use as drinking water. There is some concern that, if people drink water containing a lot of nitrate, they may become ill.

16.8 The thick, green water here is a sign of eutrophication. The green colour is caused by small photosynthetic organisms which thrive in the high concentration of nutrients in the water. Eventually, large populations of bacteria may develop, using up the oxygen in the water and stopping most other organisms from being able to live there.

Water with few nutrients is rich in oxygen, and supports a variety of animal life.

Sunlight can penetrate deep into the water, allowing water plants to grow.

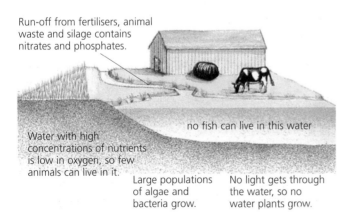

Run-off from fertilisers, animal waste and silage contains nitrates and phosphates.

Water with high concentrations of nutrients is low in oxygen, so few animals can live in it.

no fish can live in this water

Large populations of algae and bacteria grow.

No light gets through the water, so no water plants grow.

16.9 Eutrophication. Nutrients flowing into the water increase algal and bacterial growth. This reduces the oxygen levels, killing fish.

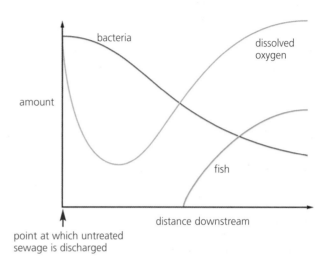

16.10 The effect of raw sewage on a stream.

Could we stop using nitrate fertilisers? It is not really sensible at the moment to suggest that we could. People expect to have plentiful supplies of relatively cheap food. Although fertilisers are expensive, by using them farmers get so much higher yields that they make more profit. If they did not use fertilisers at all, their yields would be much lower and they would have to sell their crops for a higher price, in order to make any profit at all.

Some farmers are doing just this. They do not use **inorganic** fertilisers, such as ammonium nitrate, at all. Instead, they use **organic** fertilisers, such as manure. Organic fertilisers are better than inorganic ones in that they do not contain many nitrates which can easily be leached out of the soil. Instead, they release their nutrients gradually, over a long period of time, giving crops time to absorb them efficiently. Nevertheless, manures *can* cause pollution, if a lot is put onto a field at once, at a time of year when there is a lot of rain or when crops are not growing and cannot absorb the nutrients from them.

The yields obtained when using organic fertilisers are not usually as great as when using inorganic ones, so the crops are usually sold for a higher price. Many people are now prepared to pay this extra money for food from crops grown in this way, but many cannot afford to.

If nitrate fertilisers *are* used, there is much which can be done to limit the harm they do. Care must be taken not to use too much, but only to apply an amount which the plants can take up straight away. Fertilisers should not be applied to empty fields, but only when plants are growing. They should not be applied just before rain is forecast. They should not be sprayed near to streams and rivers.

16.16 Sewage can be treated before releasing into streams.

As well as eutrophication, untreated sewage can cause other problems. It contains urine and faeces, both of which may contain harmful bacteria. If a person swims in, or drinks, water contaminated with untreated sewage, they run the risk of catching a wide variety of diseases, some of which – such as poliomyelitis – are very serious.

In many countries, most sewage is now treated to remove these harmful bacteria, and most of the nutrients which could cause eutrophication, before it is released into rivers. Sewage treatment is described in Section 14.6.

16.17 Chemical waste may contain toxins.

A very different kind of water pollution may result from the discharge of chemical waste into waterways. Chemical waste may contain **heavy metals**, such as lead, cadmium or mercury. These substances are very poisonous (toxic) to living organisms, because heavy metals stop enzymes from working. If they get into streams, rivers or the sea, they may kill almost every living thing in that area of water.

> **Questions**
> 1 Explain what is meant by 'eutrophication'.
> 2 Explain how each of the following can reduce water pollution:
>   (a) treating sewage
>   (b) using organic fertilisers rather than inorganic ones.

Pesticides

16.18 Pesticides help to increase crop yields.

A pesticide is a substance which kills organisms which damage crops. Insects which eat crops can be killed with **insecticides**. Fungi which grow on crops are controlled with **fungicides**. Weeds which compete with crop plants for water, light and minerals can be controlled with **herbicides**. Pesticides may also be used to control organisms which transmit disease, such as mosquitoes (Section 13.8).

In a natural ecosystem, a wide variety of plant species will probably grow in a particular habitat. A wide variety of animals will live in that habitat too, feeding on different plants and on each other. A natural ecosystem often has a large species diversity. Factors such as predation or food supply will prevent the population of any one species from growing too large (Sections 15.20–15.21)

On a farm, only a few plant species are allowed to grow in many of the fields. One field might contain nothing but wheat, for example. This is called a **monoculture**. Insects or fungi which can feed on the wheat have an almost inexhaustible food supply. The usual limits on their population growth do not apply. The populations of the insects or fungi may grow very rapidly, until they are so big that they cause extensive damage to the crop.

If nothing is done about this, then crop yields can be very badly reduced. It has been estimated that, in developing countries, at least one third of potential crops are destroyed by pests. If they did not use pesticides, then this would be even worse.

16.19 Pesticides can harm the environment.

By definition, a pesticide is a harmful substance. If they are not used with care, some of the pesticides can do a lot of damage to the environment. For example, **DDT** is a pesticide that kills insects. It is a **persistent** insecticide, which means that it does not break down, but remains in the bodies of the insects or in the soil. When a bird or other organism eats the insects, they eat the DDT too. The DDT stays in their bodies; each time they eat an insect, more DDT accumulates in their tissues. If a bird of prey eats the insect-eating bird, it too begins to accumulate DDT. Birds and other animals near the end of food chains can build up very large concentrations of DDT in their bodies.

Unfortunately, as well as being persistent, DDT is also **nonspecific**. This means that it not only harms the insects it is meant to kill, but is also harmful to other living things. In high concentrations it is very harmful to birds, for example. In Britain, it affected the breeding success of peregrine falcons, by making their egg shells very weak, so that they very rarely hatched. The peregrine falcon population dropped very rapidly.

Once it was realised that DDT was doing so much harm, its use in Britain was stopped. Now DDT is not used in many parts of the world. However, it is still used in many developing countries, because without it, insects would be such a problem that more people would starve or die of diseases like malaria. Other insecticides need to be developed which are as cheap and effective as DDT, but that do not harm other living organisms.

> **Questions**
> 1 What is a pesticide?
> 2 Why do we use pesticides?
> 3 Explain how pesticides can damage the environment.

An investigation in the 1970s showed that, although the DDT levels in the water of an estuary were only 0.00005 parts per million (ppm), the amounts in the animals feeding in the estuary were much greater.

protoctists 0.04 ppm → shrimps 0.16 ppm → minnows 0.50 ppm → cormorants 26.40 ppm

The numbers show the amount of DDT in each kind of animal in this food chain.

16.11 DDT accumulates along a food chain.

Conservation

16.20 Biodiversity should be conserved.

Conservation is the process of looking after the natural environment. Conservation attempts to maintain or increase the range of different species living in an area, known as **biodiversity**.

One of the greatest threats to biodiversity is the loss of habitats. Each species of living organism is adapted to live in a particular habitat. If this habitat is destroyed, then the species may have nowhere else to live, and will become extinct.

Tropical rainforests have a very high biodiversity compared with almost anywhere else in the world. This is one of the main reasons why people think that conserving them is so important. When tropical rainforests are cut down or burnt, the habitats of thousands of different species are destroyed.

We have already seen how and why deforestation occurs. Another kind of habitat that is under threat is wetland, such as swamps. People drain wetland so that it can be more easily farmed. We build roads and houses, destroying whatever used to grow on that land. We farm animals in large numbers on land that cannot really produce enough vegetation to support them, so that the land becomes a semi-desert.

Many governments and also world-wide organisations such as the World Wide Fund for Nature are aware of these problems and are attempting to make sure that especially important habitats are not damaged. Most countries have special areas where people's activities are carefully controlled, ensuring that wildlife can continue to live there. Often, the loss of money from agriculture in these areas can be regained by allowing tourists to visit them. The most successful projects actively involve local people, who are usually delighted to see their environment being cared for, so long as they can still make a living from it.

16.21 Timber production can be sustainable.

A major threat to all kinds of forest, not only tropical rainforests, comes from logging companies who cut down huge areas of trees. The wood has many uses; for example, for building, making furniture and making paper. Tropical hardwoods fetch especially high prices.

If logging of forests cannot be entirely stopped, then at least we should try to limit its damaging effects. For example, in Europe a system of woodland management called **coppicing** has been used for hundreds of years. Trees are cut down to just above ground level, and then left to regrow. This can be done every 10 or 15 years, repeatedly harvesting wood from the same trees, which just keep on regrowing. Often, a wood will be divided up into several different areas, and each year the trees in just one area are coppiced. This means that, at any one time, different parts of the wood contain trees of different sizes. This provides different habitats that suit many different species, so that biodiversity may be even higher than it would be if the wood was not harvested at all.

The trouble with coppicing is that the wood you harvest is of quite small diameter. It is not much use for building, though it can be used to make furniture or to make charcoal. To get larger pieces of timber, you need to cut down large, mature trees. If all the trees in one area are cut down, this is known as **clear-felling**. Clear-felling is harmful to the environment, because it completely destroys the forest habitat. It also leaves the soil open to erosion by rain, especially if the area is on a slope. A better method is **selective felling**, in which particular trees are cut down but others growing around them are left to grow for a few more years. However, even selective felling can cause considerable loss of biodiversity, because it usually involves heavy machinery and the construction of roads into the forest, both of which can cause considerable damage to the habitat.

16.22 Fisheries need careful management.

World-wide, most of our food comes from agriculture. However, we still consume large quantities of wild fish that we harvest from the sea. In recent years, there has been increasing concern that we are taking too many fish from the sea, reducing their populations to such small sizes that there is a risk that some species may become extinct.

An area of the sea where people catch fish on a large scale is known as a **fishery**. Fisheries exist in the seas and oceans all over the world. Many countries try to regulate the quantities of fish that can be caught in the fisheries that lie close to their shores. For example, the European Union imposes regulations on all the European countries that fish in the North Sea and North Atlantic Ocean. Each country is allocated a particular quantity of fish of each species that it is allowed to catch. There are also restrictions on the size of net that can be used, so that small fish escape and are given time to grow to adulthood and breed. Sometimes, total bans are imposed on fishing in a particular area that is known to be a breeding ground for a particular fish species, to give that species a chance to produce offspring and build up the size of its population.

These regulations are not easy to design or to enforce. Each country is always worried that another country is getting more than its fair share of fish. Fishermen in every country are worried that, if they are only allowed to catch a limited amount of fish, they will not be able to make a living. It is also difficult for scientists to find out exactly how many fish there are in a particular area, so that they can work out how many can be caught without endangering the population. So, despite the introduction of regulations by the EU in 1983, in 2001 the populations of fish such as cod in the North Sea were continuing to get smaller and smaller. If we do not want to lose many different species of fish from the seas, then fishery regulations must become even tougher.

16.23 Natural resources should be conserved.

As well as trees, fish and other living organisms, we need to be aware of other natural resources that we use, and attempt to conserve them. For example, we use very large amounts of fossil fuels such as coal and oil, that are extracted from the ground. They are burnt to release energy that is used to generate electricity. Oil is also used to make other substances, such as plastics.

These fossil fuels formed in the Earth many millions of years ago. If we keep on using them, they will eventually run out. Moreover, we have seen how burning fossil fuels can produce acid rain (Sections 16.6–16.8) and can contribute to global warming (Section 16.3).

For all of these reasons, it is important that we should try to reduce our use of these limited resources. We can do this by reducing the amount of energy that we use, and also looking for other sources of renewable energy, such as wind, water or solar power.

16.24 Recycling can help to conserve resources.

Resources can be conserved if, instead of throwing things away when we have used them, we recycle them so that they can be used again.

We have already seen how water – which is a scarce resource in many parts of the world – can be recycled (Sections 14.6–14.7). Another substance that is often recycled is paper. Paper is made from trees, so if used paper can be pulped and used to make paper again, this should reduce the number of trees that are cut down. However, things are not quite so straightforward. The trees that are used to make paper are often especially planted, and when they are cut down new ones are planted to take their place. So paper-making need not actually threaten natural forests. Another problem is that collecting and transporting waste paper so that it can be recycled uses a lot of energy, so the paper that is made may end up costing more than 'first time around' paper.

Glass can also be recycled. Glass is made from silicon oxide, usually in the form of sand. In many countries, used glass bottles are collected and taken to recycling plants. The glass is melted down and used to make new bottles and other glass objects. If this is to work well, then different kinds of glass – clear, green and brown – need to be collected separately. Once again, transport costs have to be taken into account. The process of melting and reforming the glass also uses a lot of energy. A much better way of recycling glass is to reuse glass bottles, and this is done in an increasing number of countries around the world.

Some plastics can also be recycled. Many plastics are non-biodegradable – they do not rot and so remain as unsightly rubbish when discarded. Recycling plastics reduces this pollution problem, and also helps to conserve the fossil fuels from which many plastics are made.

Questions

1. Explain the meaning of (a) biodiversity and (b) conservation.
2. Explain the advantages of (a) coppicing and (b) selective felling in the conservation of forests.
3. List three measures that can be taken to ensure that fish populations can continue to be fished for many years in the future.
4. For one named resource, explain the advantages and problems of recycling it.

17 The diversity of life

17.1 Classification involves grouping things.

Classification means putting things into groups. Biologists classify living organisms according to how closely they think they are related to one another.

17.2 We use a classification system for living things.

The first person to try to classify living things in a scientific way was a Swedish naturalist called **Linnaeus.** He introduced his system of classification in 1735. He divided all the different kinds of living things into groups called **species**. He recognised 12 000 species. Linnaeus' species were groups of organisms which had a lot of features in common. We still use this system today.

Species are grouped into **genera** (singular: **genus**). Each genus contains several species with similar characteristics (Fig 17.1). Several genera are then grouped into a **family**, families into **orders**, orders into **classes**, classes into **phyla** and finally phyla into **kingdoms**. Some of the more important ones are described in this chapter.

17.3 Each species has two Latin names.

Linnaeus gave every living organism two names, written in Latin. The first name is the name of the genus it belongs to, and always has a capital letter. The second name is the name of its species, and always has a small letter.

For example, a wolf belongs to the genus *Canis* and the species *lupus*. Its Latin name is *Canis lupus*. These names are printed in italics. When you write one, you cannot really write in italics, so you should underline any Latin names. The genus name can be abbreviated like this: *C. lupus*.

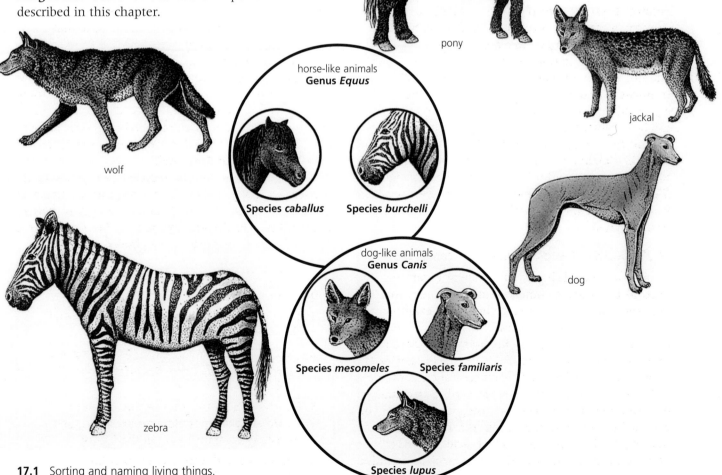

17.1 Sorting and naming living things.

Kingdom Bacteria

These are the bacteria. Some can carry out photosynthesis. The oldest fossils belong to this kingdom, so we think that they were the first kind of organisms to evolve.

Characteristics
often unicellular (single-celled) organisms,
have no nucleus,
have a cell wall.

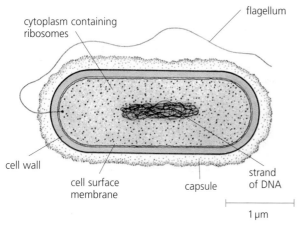

17.2 Cross-section of a bacterium, *Escherichia coli*.

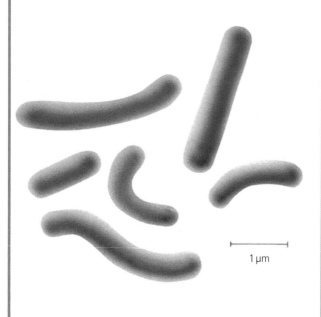

17.3 External view of the bacterium, *Vibrio cholerae* (cholera bacterium).

Kingdom Fungi

Fungi do not have chlorophyll, and do not photosynthesise. Instead they feed saprophytically, or parasitically, on organic material like faeces, human foods and dead plants or animals.

Characteristics
multicellular (many-celled),
cells have nuclei,
have cell walls,
do not have chlorophyll,
feed by saprophytic or parasitic nutrition.

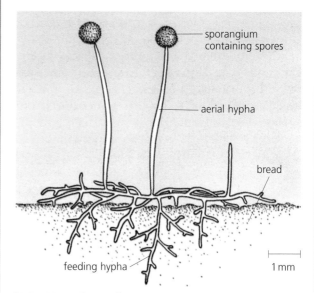

17.4 *Mucor haemalis*.

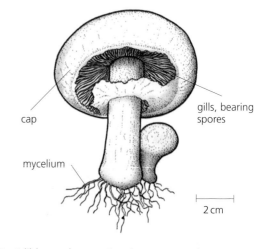

17.5 Edible mushroom *Agaricus campestris*.

Kingdom Plantae

Almost all plants are green because at least some of their cells contain chloroplasts. The chloroplasts contain chlorophyll, which absorbs sunlight for photosynthesis.

Characteristics
multicellular, have cell walls,

Division Angiospermophyta

These are the flowering plants (angiosperms).

Characteristics
have roots, stems and leaves,
have xylem and phloem,
reproduce by producing seeds,
seeds produced inside ovary, inside flower.

Monocotyledonous plants
strap-shaped leaves with parallel veins,
one cotyledon in a seed.

Dicotyledonous plants
leaves which can be broad, and which have a network of branching veins,
two cotyledons in a seed.

17.6 Corn, *Zea mays*, a monocot.

17.7 Pea, *Pisum sativum*, a dicot.

Kingdom Animalia

Animals do not photosynthesise, so they never have chlorophyll. They eat other living organisms, so they are usually able to move, to find their food. They do not have cell walls, as this would make it difficult for them to move easily.

Characteristics
multicellular,
do not have cell walls,
do not have chlorophyll,
feed heterotrophically.

Phylum Annelida

These are worms, with bodies made up of ring-like segments. Most of them live in water, though some, like the earthworm, live in moist soil.

Characteristics
animals with bodies made up of ring-like segments,
no legs,

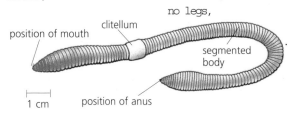

17.8 Earthworm, *Lumbricus terrestris*.

Phylum Mollusca

These are soft-bodied animals, sometimes with a shell, like snails, or without, like slugs. Octopuses are also molluscs.

Characteristics
animals with soft, unsegmented bodies,
may have shell.

17.9 Snail.

Phylum Nematoda

Nematodes are worms, but unlike annelids their bodies are not divided into segments. They are usually white, long and thin. They live in many different habitats. Many nematodes live in the soil.

Characteristics
animals with long, thin bodies with no rings, no legs,

17.10 Nematode.

Phylum Arthropoda

Arthropods are animals with jointed legs, but no backbone. They are a very successful group, because they have a waterproof exoskeleton that has allowed them to live on dry land. There are more kinds of arthropod in the world than all the other kinds of animals put together.

Characteristics
animals with several pairs of jointed legs, have an exoskeleton.

Class Crustacea

These are the crabs, lobsters and woodlice. They breathe through gills, so most of them live in wet places and many are aquatic.

Characteristics
arthropods with more than four pairs of jointed legs, breathe through gills,
two pairs of antennae.

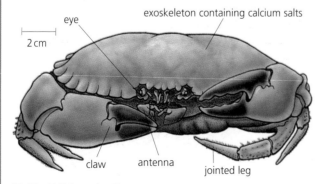

17.11 Edible crab, *Cancer pagurus*.

Class Arachnida

These are the spiders, ticks and scorpions.

Characteristics
arthropods with four pairs of jointed legs, breathe through gills called book lungs.

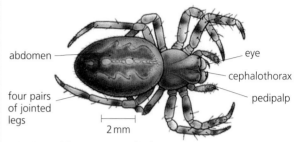

17.12 Spider, *Areneus diadema*.

Class Insecta

Insects are a very successful group of animals. Their success is mostly due to their exoskeleton and
tracheae, which are very good at stopping water from evaporating from the insects' bodies, so they can live in very dry places. They are mainly terrestrial.

Characteristics
arthropods with three pairs of jointed legs,

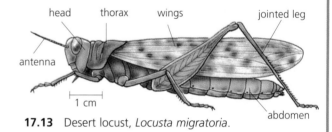

17.13 Desert locust, *Locusta migratoria*.

Class Myriapoda

These are the centipedes and millipedes. Centipedes are fast-moving carnivores, while millipedes are
herbivores.

Characteristics
arthropods, with a pair of legs on each segment.

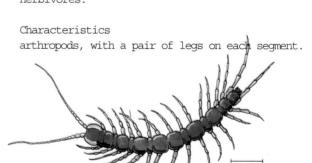

17.14 Centipede, *Lithobius forficatus*.

Phylum Chordata

These are animals with a supporting rod running along the length of the body. The most familiar ones have a backbone, and are called vertebrates.

Class Pisces

The fish all live in water, except for one or two like the mudskipper, which can spend short periods of time breathing air.

Characteristics
vertebrates with scaly skin,
have gills,
have fins.

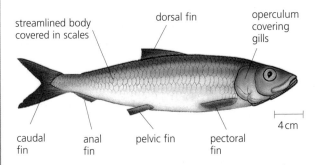

17.15 Fish, *Clupea harengus* (herring).

Class Amphibia

Although most adult amphibians live on land, they always go back to the water to breed. Frogs and toads are amphibians.

Characteristics
vertebrates with moist, scale-less skin,
eggs laid in water, larva (tadpole) lives in water, but adult often lives on land,
larva has gills, adult has lungs.

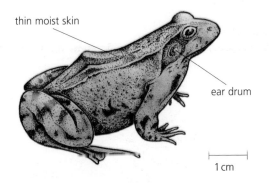

17.16 Frog, *Rana temporaria*.

Class Reptilia

These are the crocodiles, lizards, snakes, turtles and tortoises. Reptiles do not need to go back to the water to breed because their eggs have a waterproof shell which stops them from drying out.

Characteristics
vertebrates with scaly skin,

17.17 Snake, *Natrix natrix*.

Class Aves

The birds, like reptiles, lay eggs with waterproof shells.

Characteristics
vertebrates with feathers,
forelimbs have become wings,
lay eggs with hard shells,
homeothermic,
have a beak.

17.18 Broad-winged hawk, *Buteo platypterus*.

Class Mammalia

This is the group that humans belong to.

Characteristics
vertebrates with hair,
have a placenta,
young fed on milk from mammary glands,
homeothermic,
have a diaphragm,
heart has four chambers,
have different types of teeth (incisors, canines, premolars and molars),
cerebral hemispheres very well developed.

17.19 Ocelot, *Felis pardalis*.

Apparatus required for practicals

The apparatus listed is that required for each group performing the experiment.

1.1 Looking at animal cells

Follow local recommendations about safety aspects of using cheek cells.
section lifter
slide
coverslip
pipette
very small amount of methylene blue (diluted)
filter paper or blotting paper
microscope

1.2 Looking at plant cells

Sections of onion bulb; these can be cut beforehand and kept in a beaker of water
slide
coverslip
pipette
filter paper or blotting paper
microscope
seeker, needle or forceps

2.1 To show diffusion in a solution

gas jar
crystal of potassium permanganate

2.2 To find the effects of different solutions on plant cells

solution A – distilled water
solution B – $0.3\,mol\,dm^{-3}$ sucrose solution
solution C – $1.0\,mol\,dm^{-3}$ sucrose solution
red rhubarb petioles or other coloured plant tissue
forceps
scalpel
pipette
3 microscope slides and coverslips
labels for slides
filter paper or blotting paper
microscope

2.3 To demonstrate osmosis using eggs

Session 1:
2 fresh eggs
2 large beakers containing enough dilute HCl to cover eggs
Session 2:
de-shelled eggs
two large beakers
Chinagraph pencil or glass marker pen
distilled water
20% salt solution

2.4 The effect of osmosis on potato tissues

raw potatoes
5 beakers or other containers
cork borer and/or knife
access to sucrose solutions of approx. 0.1, 0.25, 0.5 and $1.0\,mol\,dm^{-3}$
access to water
ruler to measure in mm

3.1 The effect of catalase on hydrogen peroxide

safety goggles
raw and boiled potato
yeast suspension, produced by mixing a little dried yeast and a little sucrose into warm water
chopped fresh liver
fresh fruit juice
dropper pipettes or other means of transferring liquids into tubes
syringe, pipette or measuring cylinder for measuring $10\,cm^3$
about $50\,cm^3$ of 10 volume or 20 volume hydrogen peroxide per group
5 boiling tubes and rack
wooden splint and access to lit Bunsen flame

3.2 Investigating the effect of temperature on enzyme activity

4 test tubes and rack
beakers
1% amylase solution
water baths at 35°C and 80°C
access to refrigerator
thermometer
boiling tube
10% starch solution
iodine solution
spotting tile
4 glass rods
stop watch
Chinagraph pencil or glass marker pen
pipette
2 syringes to measure $5\,cm^3$

3.3 Investigating biological washing powders

It is not possible to give a detailed apparatus list for this practical, as students' designs are likely to vary widely.
Notes:
It is easier to obtain quantitative results using developed film rather than using proteinaceous stains on cloth. However, students will much prefer to use 'real' stains on 'real' fabrics. Stains can be made using egg or 'blood' (in reality myoglobin) from fresh meat, but teachers would be well advised to test how such stains perform beforehand. The best types of fabric to use are either cotton or polyester cotton.
For safety reasons, teachers may prefer to use encapsulated proteases and lipases. These can be obtained from the following address:
National Centre for Biotechnology Education,
Department of Microbiology,
University of Reading,
Whiteknights,
Reading,
RG6 2AJ
England.

A washing powder enzyme pack, containing five different enzymes, can be obtained from the same address.
Suitable amounts of enzyme to use are in the range of 1 to 2g of enzyme for every 100g of washing powder. 1g of this mixture should then be dissolved in $100\,cm^3$ of water.

3.4 Investigating the use of pectinase in making fruit juice

It is not possible to give a detailed apparatus list for this practical, as student's designs will vary widely.
Pectinase and fruit from different sources will require different proportions of pectinase to fruit. However, as a general guide, about 1g of pectinase to 100g of crushed fruit will give reasonably rapid results. This is a much higher amount than would be used industrially – another reason why the fruit juice extracted should not be consumed.

4.1 Testing food for carbohydrates

variety of foods
glucose solution
sucrose solution
starch power
test tube rack
boiling tubes
Bunsen burner and tripod
beaker
test tube holder
pipette
white tile
scalpel
Benedict's solution
iodine solution

4.2 Testing food for proteins

variety of foods
test tube rack
test tubes
pipettes
tile
scalpel
potassium hydroxide solution and 1% copper sulphate solution or biuret solution

4.3 Testing food for fats

variety of foods
cooking oil
test tube rack
clean, dry test tubes
pipette
tile
scalpel
absolute alcohol
distilled water

5.1 Looking at the epidermis of a leaf
variety of leaves
forceps
slides and coverslips
microscope
pipette
clear nail varnish

5.2 Testing a leaf for starch
Pelargonium or *Coleus* plant which has been photosynthesising
boiling water bath, or beaker etc.
boiling tube
methylated spirits
glass rod
iodine solution
forceps
white tile
pipette

5.3 To see if light is necessary for photosynthesis
Session 1:
Pelargonium or *Coleus* plant
Session 2:
destarched plant
black paper or aluminium foil
scissors
paperclips
Session 3:
apparatus as for 5.2

5.4 To see if carbon dioxide is necessary for photosynthesis
Session 1:
Pelargonium or *Coleus* plant
Session 2:
destarched plant
2 conical flasks, fitted with split corks
potassium hydroxide solution
distilled water
petroleum jelly
clamp stands or other means of support for flasks
Session 3:
apparatus as for 5.2

5.5 To see if chlorophyll is necessary for photosynthesis
Session 1:
plant with variegated leaves
Session 2:
destarched plant with variegated leaves
apparatus as for 5.2

5.6 To show that oxygen is produced during photosynthesis
Session 1:
large beaker
funnel which fits entirely inside beaker
test tube
Canadian pondweed (*Elodea*) or other
Session 2:
Bunsen burner
wooden splint

6.1 To show that peanuts release energy when they are oxidised
Ensure no students have an allergy to nuts before attempting this practical.
Bunsen burner
heat-proof mat
a peanut (ground nut)
a mounted needle
clamp stand, clamp and boss
boiling or test tube
thermometer

6.2 To show that heat is produced in respiration
Session 1:
pea seeds
2 beakers and Bunsen burner
Session 2:
boiled peas from Session 1
soaked peas from Session 1
mild disinfectant solution
2 vacuum flasks
2 cotton wool plugs to fit flasks tightly
2 thermometers
clamp stands to support flasks

6.3 To show that carbon dioxide is produced when yeast respires anaerobically
4 boiling tubes, fitted with bungs and glass tubing as in Fig 6.3
boiled, cooled water, or apparatus for boiling it
sucrose or glucose
fresh or dried yeast
boiled yeast solution
glass rod
pipette
lime water or hydrogencarbonate indicator
two beakers to support boiling tubes
Chinagraph pencil or glass marker pen
liquid paraffin

6.4 Examining lungs
Follow local recommendations about safety aspects of using animal organs.
set of sheep's or cow's lungs
burette tube

6.5 Comparing the carbon dioxide content of inspired and expired air
see Fig 6.12

6.6 Investigating how breathing rate changes with exercise
stop watch
A good exercise in a confined space is stepping on and off a chair.

7.1 To find the effect of exercise on the rate of heart beat
stop watch

7.2 To see which part of a stem transports water and solutes
Session 1:
whole, small plant (for example, a freshly pulled weed)
eosin solution
beaker
Session 2:
plant from Session 1
slide
cover slip
razor blade
tile
paint brush or section lifter
pipette
microscope

7.3 To see which surface of a leaf loses most water
potted plant with smooth leaves
forceps
cobalt chloride paper in desiccator
self-adhesive book-covering film
scissors

7.4 To measure the rate of transpiration of a potted plant
2 plants of similar size, in pots of the same size
2 large polythene bags
rubber bands
petroleum jelly
balance

7.5 Using a potometer to compare rates of transpiration under different conditions
a potometer
plant with firm stems, such as geranium, which will fit tightly into the apparatus
wire
pliers
petroleum jelly
stop watch
electric fan
access to a refrigerator

8.1 Investigating the structure of an insect-pollinated flower

flower stalk of an insect-pollinated flower, with flowers in various stages of development
hand lens
razor blade
tile
microscope slide
microscope
seeker

8.2 Growing pollen tubes

4 cavity slides
petroleum jelly
4 coverslips
labels for slides
variety of sugar solutions, e.g. distilled water, $0.1\,mol\,dm^{-3}$ sucrose, $0.5\,mol\,dm^{-3}$ sucrose, $1.0\,mol\,dm^{-3}$ sucrose, each with a very small amount of boric acid added
4 types of flowers, with ripe pollen
seeker
microscope
incubator at $20\,°C$

8.3 To find the conditions necessary for the germination of mustard seeds

5 test tubes, fitted with gauze or perforated zinc platforms
pyrogallol in NaOH solution (take care, this is very caustic)
cotton wool
rubber bung to fit test tube
test tube racks
mustard seeds
Chinagraph pencil or glass marker pen

9.1 Looking at human eyes

small mirror

9.2 Dissecting a sheep's eye

Follow local recommendations about safety aspects of handling eyes.
a sheep, bullock, or pig eye
dissecting board or dish
forceps, scalpel and scissors
newspaper
paper towels

9.3 Measuring reaction time

stop watch

9.4 To find out how shoots respond to light

3 petri dishes
Chinagraph pencil or glass marker pen
cotton wool or filter paper
mustard seeds
2 light-proof boxes, one with a slit in one end
clinostat

9.5 To find out how roots respond to gravity

Session 1:
broad bean seeds
blotting paper
gas jars
Session 2:
2 clinostats
blotting paper
pins

10.1 Investigating the effect of size and covering on rate of cooling

beaker
Bunsen burner, tripod, heatproof mat and gauze
paper towel or cloth for holding hot beaker
enough cotton wool to cover a boiling tube
rubber band
test tube rack to hold boiling tubes
small test tube
2 large boiling tubes
Chinagraph pencil or glass marker pen
thermometer
stop clock
corks to fit each tube

10.2 Investigating the effect of evaporation on rate of cooling

It is quite possible to include this investigation in Practical 10.1, simply by adding another boiling tube which is wrapped in wet cotton wool. However, this introduces a lot of ideas simultaneously, and, unless the students are of high ability or are already familiar with these ideas, it is probably best to separate them.
2 retort stands, clamps and bosses
2 thermometers
cotton wool
rubber bands
stop clock

12.1 Breeding beads

two containers
150 beads of one colour (red)
50 beads of a second colour (yellow)

14.1 Growing bacteria and fungi on agar jelly

sterile Petri dishes, ready poured with nutrient agar
Chinagraph pencil or glass marker pen
inoculating loop, and access to Bunsen flame
pond water in small container
adhesive tape
incubator at about $25\,°C$

14.2 What factors affect the way that dough rises

For the basic bread dough:
a little sugar
a little dried yeast
about $75\,g$ of strong flour
a beaker or other container for mixing

To measure how fast or how much the dough rises, it can be left to rise in a transparent measuring cylinder.

For investigating the factors suggested, the following materials will be needed:
thermometer
sodium chloride
different types of flour, e.g. plain, strong, white, brown, rye, etc.
different types of yeast, e.g. fresh, dried, brewers etc.
ascorbic acid (vitamin C)
amylase

The time taken for the dough to rise varies considerably; this is largely dependent on the activity of the yeast used, and it is well worth finding a good source of active yeast in order to speed up the results from this investigation. Students will almost certainly need to visit their dough some time after their lesson has ended, in order to measure its volume.

14.3 Making yoghurt

a starter culture of 'live' yoghurt – plain live, unpasteurised yoghurt can be bought in large supermarkets and health food shops
sterile containers in which the yoghurt can be made, for example test tubes
sterile syringe or small measuring cylinder for measuring $10\,cm^3$ – alternatively, a line can be drawn on the test tubes at a level approximating to this volume
pasteurised milk
cling film
water bath or incubator at about $40\,°C$

For investigating the factors suggested, the following materials will be needed:
Universal Indicator paper
lactase
different types of milk, e.g. raw, sterilised, UHT, goat's
different types of starter culture
water baths or incubators at different temperatures

Examination questions

1 Two characteristics of living organisms are nutrition and respiration.

(a) (i) List **three** other characterisitics of living organisms. [3]

(ii) Name the process by which green plants produce carbohydrates. [1]

(b) Living organisms release gases into the atmosphere as a result of their various activities. Copy and complete the table, using a tick (✓) or a cross (✗), to show which gases are released.

	carbon dioxide released into the atmosphere	oxygen released into the atmosphere
animals in bright light		
green plants in bright light		
animals in the dark		
green plants in the dark		

[4]

0610/2 J00   Total [8]

2 A student who had seen a cook place slices of fresh pineapple on top of meat before cooking it, decided to carry out some experiments using fresh pineapple juice.

A milk agar was prepared by mixing fat-free milk and warm water with agar powder. It was then poured into a sterilised petri dish and allowed to cool. After it had set, five holes were made in the milk agar. Each hole was filled with a different substance, as shown in Fig. 2.1.

Key
1 – distilled water
2 – a solution of lactase (an enzyme which digests milk sugar)
4 – boiled pineapple juice
5 – fresh pineapple juice

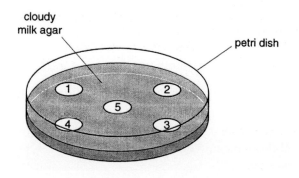

Fig 2.1

The petri dish was then kept at 40°C for 2 hours. The results obtained are shown in Fig 2.2.

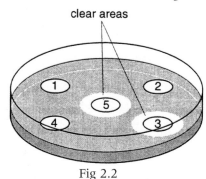

Fig 2.2

(a) Explain why the petri dish was kept at 40°C. [2]
(b) Explain the results obtained for each of the holes 1–5. [5]
(c) Using the results of this experiment, suggest reasons why cooks place fresh pineapple on meat before cooking it. [2]

5090/2 N00   Total [9]

3 Fig 3.1 shows a root hair cell.

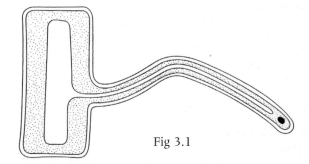

Fig 3.1

(a) Label (i) the **cell membrane** and (ii) the **vacuole**. [2]
(b) State **two** differences in structure between this cell and a palisade cell from a leaf. [2]
(c) One function of the root hair cell is the absorption of water.
   (i) How is the structure of the root hair cell adapted to this function? [1]
   (ii) State one other function of this cell. [1]
(d) Water is taken in to the cell by osmosis. Define **osmosis**. [3]
(e) Water is passed to the centre of the root and then along it and up the stem.
   (i) In which tissue does water travel through the plant? [1]
   (ii) State **two** uses for the water taken in by a plant. [2]

0610/2 N00   Total [12]

4 Fig 4.1 shows an experiment to investigate the uptake of ions by a plant.

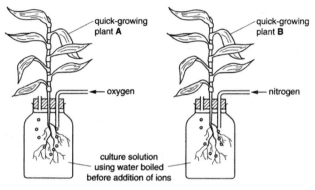

Fig 4.1

The culture solutions contained measured quantities of all the ions necessary for plant growth, dissolved in water which had been boiled to remove dissolved gases.

(a) Name the process, characteristic of all plant cells, which the roots of plant **B** will be unable to carry out. [1]

Using the radioactive form of an ion, the rate at which it is absorbed from culture solutions can be measured. Fig 4.2 shows the rate of uptake of one particular ion from the two solutions in Fig 4.1.

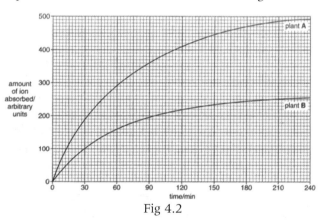

Fig 4.2

(b) Using Fig 4.2,
(i) state how long it took plant **A** to absorb 330 units of the ion;
(ii) calculate the difference in intake of the ion between the two plants after **three** hours. [2]
(c) Explain the difference in ion uptake between the two plants. [3]
(d) Describe and explain the appearance of plant **A** if the culture solution had been deficient in
(i) magnesium
(ii) nitrogen [3]

5090/2 N99   Total [9]

5 Fig 5.1 shows a blood vessel (**Q**) linking a part of the alimentary canal (**P**) with an organ (**R**) in the abdomen.

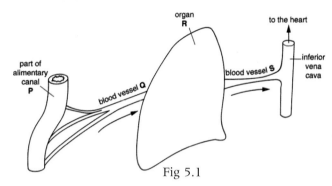

Fig 5.1

(a) (i) Identify **P**, **Q** and **R**. [3]
(ii) What type of blood vessel is **S**? [1]
(b) State the changes in the composition of the blood in **Q** shortly after a meal has been eaten which contains protein and carbohydrate. [3]

An athlete is about to take part in a race.

(c) (i) How do the concentrations of materials in blood vessel **S** differ from those in blood vessel **Q**? [2]
(ii) Explain how the differences in concentration of the materials have occurred. [2]

0610/3 J96   Total [11]

6 Fig 6.1 shows the volumes of air exchanged by a student at rest and then during vigorous exercise.

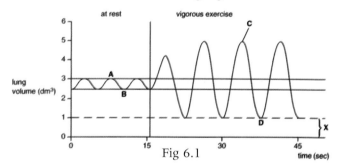

Fig 6.1

(a) Using Fig 6.1, calculate
(i) the volume of air exchanged per minute at rest;
(ii) the largest volume of air exchanged in one breath during exercise. [2]
(b) Which letter, **A**, **B**, **C** or **D** on the graph represents the maximum contraction of the
(i) diaphragm muscle
(ii) internal intercostal muscles? [2]
(c) The alveoli in people who smoke many cigarettes each day, lose their elasticity. What effect will this have on volume **X**? [1]

5096/1 J00   Total [5]

7 Fig 7.1 shows two experiments to investigate the partial permeability of Visking tubing.

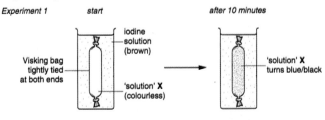

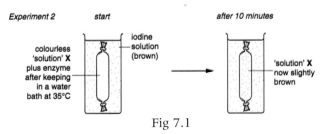

Fig 7.1

(a) Suggest what 'solution' **X** was likely to have been. [1]

(b) In Experiment 1, explain
 (i) why 'solution' **X** turned from colourless to blue/black; [2]
 (i) why the iodine solution remained brown. [1]

In Experiment 2, 'solution' **X** and an enzyme were placed in a Visking bag which was kept at 35°C for 30 minutes. After this time, the bag was placed in iodine solution. This experiment, and the results, are also shown in Fig 7.1.

(c) In Experiment 2, explain
 (i) why the bag was first kept at 35°C for 30 minutes. [1]
 (ii) why 'solution' **X** did not turn blue/black. [1]

At the end of Experiment 2, the student noticed a change in the condition of the Visking bag.

(d) (i) What change might have been noticed? [1]
 (ii) Explain what caused this change. [2]

0610/3 N98 Total [9]

8 Fig 8.1 shows an experiment on gaseous exchange in a potted plant.

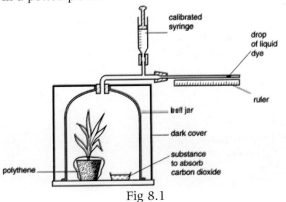

Fig 8.1

(a) Which metabolic process is being investigated in this experiment? [1]
(b) In which direction would the drop of dye move during the experiment? Explain your answer. [3]
(c) Suggest **two** uses for the syringe in this experiment. [2]

The dark cover is then removed and the plant exposed to bright light.

(d) Suggest why the drop of dye might stop moving. [2]

5090/2 J00 Total [8]

9 Fig 9.1 shows a root hair cell surrounded by soil.

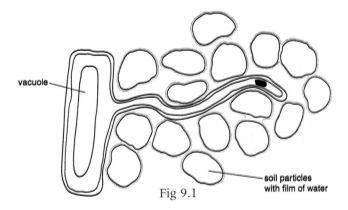

Fig 9.1

The water film around soil particles is a dilute solution of mineral salts.

(a) Explain how water passes from the soil into the vacuole of the root hair cell. [4]
(b) Mineral salts are also absorbed by the root hair cell.
 (i) In which tissue are mineral salts carried to the rest of the plant? [1]
 (ii) State **one** role of magnesium ions in a plant. [1]
(c) In an investigation, the rate of transpiration for different sizes of stomatal opening were recorded in still air conditions and in windy conditions. The results are shown below.

size of stomatal opening (μm)	rate of transpiration (arbitrary units)	
	in still air	in wind
0	0	0
5	45	95
10	55	155
15	60	210
20	65	260

 (i) Plot the results, using one pair of axes. [3]
 (ii) Suggest reasons for the shapes of the curves. [4]

0610/2 J01 Total [13]

10 Fig 10.1 shows how the amounts of different nutrients in food change as the food passes along the alimentary canal. The width of each band indicates the amount of nutrient.

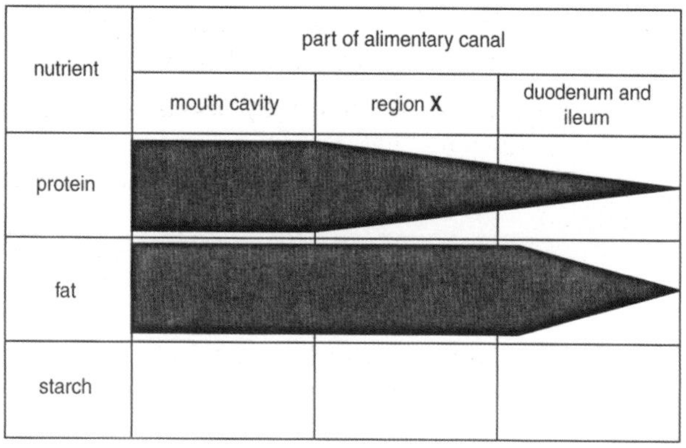

nutrient	part of alimentary canal		
	mouth cavity	region **X**	duodenum and ileum
protein			
fat			
starch			

Fig 10.1

(a) Identify region **X**. [1]
(b) Using Fig 10.1, and your own knowledge, explain what happens to protein from when it enters the mouth to when it leaves the ileum. [3]
(c) (i) Name the end-products of fat digestion. [1]
(ii) Explain why the amount of fat does not begin to decrease immediately on entering the duodenum. [2]
(d) Copy and complete Fig 10.1 to show what happens to starch from when it enters the mouth to when it leaves the ileum. [2]
5090/2 J01   Total [9]

11 (a) State **three** normal functions of a root. [3]
Fig 11.1 shows a mangrove tree growing in a swamp.

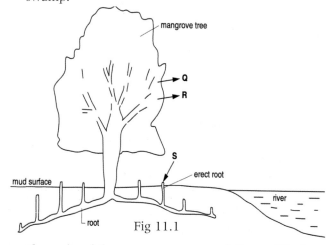

Fig 11.1

The roots of the mangrove are specially modified to overcome the fact that air spaces in the soil are always filled with water. In other respects, the roots are normal in structure and function.

Arrows **Q**, **R** and **S** represent the movement of gases in and out of the tree during the day.

(b) (i) Name gases **Q** and **R** and, for each gas, state the process in the tree which produces it. [2]
(ii) Name gas **S** and state in which process it is used in the tree. [1]
(c) Name the unusual response being shown by the erect roots of the tree. [1]

Active transport is a process which occurs in most plant roots.

(d) Suggest why mangrove roots may have difficulty in carrying out this process. [2]
0610/3 J98   Total [9]

12 Describe the relationship between the structure and function of
(a) xylem vessels; [5]
(b) the different types of blood vessel. [7]
5090/2 J01   Total [12]

13 (a) Define **asexual reproduction**. [3]
(b) Describe the process of asexual reproduction in a potato plant. [7]
(c) Potatoes can also reproduce sexually, forming seeds. Discuss the advantages **to potato plants** of asexual and sexual reproduction. [5]
0610/3 J01   Total [15]

14 (a) Describe, with the aid of diagrams, an experiment to show the environmental conditions that affect seed germination. [7]
(b) Compare the ways in which a plant embryo and a human embryo obtain the raw materials for nutrition and respiration. [5]
5090/2 J01   Total [12]

15 Distinguish clearly between the following pairs of terms.
(a) **ovary** and **ovule** [4]
(b) **ureter** and **urethra** [4]
(c) **testa** and **testis** [4]
(d) **fertilisation** and **pollination** [3]
0610/3 J96   Total [15]

16 (a) Draw labelled diagrams to show the positions of xylem and phloem tissues in
(i) a root;
(ii) a stem. [5]
(b) Describe an experiment to show which tissue in a stem conducts water from roots to leaves. [4]
(c) Explain how a plant stem is able to support the leaves, flowers and fruits. [3]
5090/2 J00   Total [12]

17 Fig 17.1 shows a section through the eye.

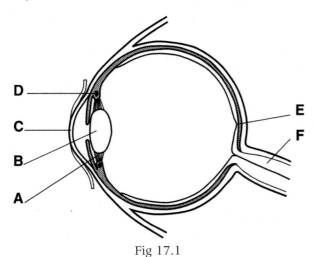

Fig 17.1

(a) Explain what happens in the eye when a person reads the words on the page of a book. Your answer should refer to, and identify, structures **A** to **F** on the diagram. [12]
(b) Suggest why it is an advantage to have two eyes instead of one. [3]
0610/3 J96   Total [15]

18 (a) Describe the difference between a taxic and a tropic response. [5]
(b) Explain how a named stimulus leads to a reflex response in a mammal. [7]
5090/2 J01   Total [12]

19 Pregnant women at high risk of having a baby with Down's syndrome are often offered an amniocentesis. This technique is shown in Fig 19.1.

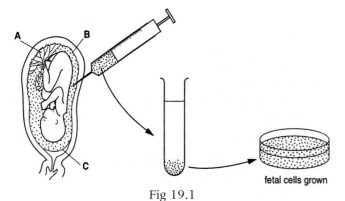

Fig 19.1

(a) Identify the parts labelled **A**, **B** and **C** and state a function of each one. [6]

The technique involves taking a sample of **B** from within the uterus. Fetal cells in the sample are then grown and analysed.

(b) (i) Suggest how the cells would be different from normal cells if the fetus has Down's syndrome. [1]
    (ii) What is the cause of this difference? [1]
(c) Suggest how the sex of the fetus could be identified by observation of fetal cells. [3]

During pregnancy women may also be monitored in other ways, including urine sampling.
(d) Suggest why the urine of pregnant women is analysed. [2]
0610/3 N00   Total [15]

20 (a) List the main parts of the human female reproductive system and describe the function of each part listed. [8]
(b) (i) Describe the signs and symptoms of a **named** sexually-transmitted disease.
    (ii) How can the spread of this disease be controlled? [4]
5090/2 J00   Total [12]

21 Fig 21.1 shows a kidney and its associated blood vessels.

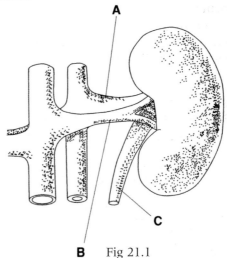

Fig 21.1

(a) (i) Name tube **C**. [1]
    (ii) Which vessel, **A** or **B**, is the renal artery? Give a reason for your choice. [2]
(b) Examine the data in Table 21.1.

	Concentration in blood plasma entering the kidney / g per 1000 cm^3	Concentration in urine / g per 1000 cm^3
water	910	960
protein	74	0
glucose	1	0
urea	0.3	20
salt (sodium chloride)	9	12

Table 21.1

What are **three** main functions of the kidney shown in the data? [3]
(c) (i) Where is the urea formed? [1]
(ii) From what is urea formed? [1]

0610/2 N95   Total [8]

22 (a) State two advantages of seed dispersal. [2]

Fig 22.1 shows two fruits, **W** and **X**, in longitudinal section.

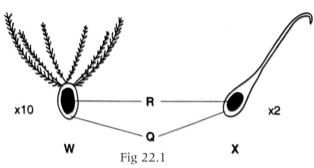

Fig 22.1

(b) (i) Name the structure **R** in the two fruits.
(ii) From which part of the parent plants have the structures labelled **Q** developed? [2]
(c) Suggest and explain how **each** of the fruits might be dispersed. [4]
(d) Suggest **three** reasons why a dispersed fruit may **not** produce a new plant. [3]

5090/2 J00   Total [11]

23 Fig 23.1 shows *Euglena gracilis*, a single-celled organism, often found in freshwater ponds.

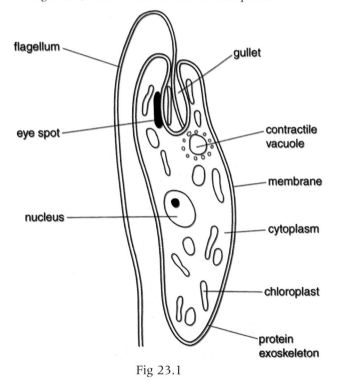

Fig 23.1

Euglena shows a number of characteristics of living things such as excretion, nutrition and irritability.
(a) Name **three** other characteristics of living things that you would expect this organism to show. [3]
(b) *Euglena* is difficult to classify because it shows animal characteristics and plant characteristics, some of which are listed in the table below. Copy the table. For each characteristic, identify it as an animal, plant or bacterial cell feature by putting a tick (✔) for present or a cross (✘) for absent in each box of your table.

feature	animal cell	plant cell	bacterial cell
chloroplast			
cytoplasm			
membrane			
nucleus			

[4]

(c) The cytoplasm of *Euglena* contains salts that are more concentrated than those in the surrounding water. The contractile vacuole excretes any excess of water.

Explain why this function of the contractile vacuole is important to this organism. [3]
(d) *Euglena* has an eye spot that is sensitive to light.
(i) Suggest and explain how the organism would respond if there was an area of brighter light nearby. [2]
(ii) Explain how the organism would benefit from this reaction. [2]

0610/3 J01   Total [14]

24 (a) Explain why osmosis may be considered to be a special type of diffusion. [3]
(b) Describe and explain the effects of placing into pure water
(i) a plant cell, and
(ii) an animal cell. [6]
(c) When the rate of water loss from a plant is greater than the rate of water uptake, the stomata close. Explain
(i) how stomata close, and
(ii) how this closing is an advantage to the plant under these conditions. [3]

5090/2 N01   Total [12]

25 (a) Using suitable examples, distinguish between **continuous** and **discontinuous variation**. [6]
(b) (i) Define **mutation**. [2]
(ii) State the factors that cause an increase in the rate of mutation and describe the effects of these mutations in humans. [4]

(iii) Explain the incidence of sickle cell anaemia in relation to that of malaria. [3]

Total [15]

26 Fig 26.1 shows a reflex arc.

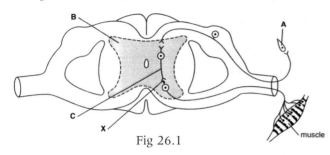

Fig 26.1

(a) (i) Name the parts labelled **A**, **B** and **C**; [3]
(ii) Copy part **C** and show, by means of an arrow, the direction in which nerve impulses pass. [1]

(b) State the name of a substance which slows down chemical transmission at **X**. [1]

Total [5]

27 Fig 27.1 shows a section through the heart.

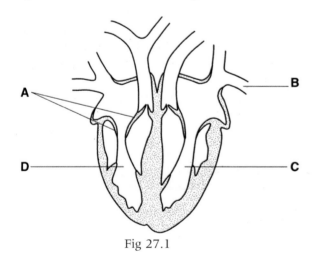

Fig 27.1

(a) Copy Fig 27.1.
(i) Name the parts labelled **A** and **B**. [2]
(ii) Shade the cavity of the ventricle which contains oxygenated blood. [1]
(ii) Suggest why the wall around chamber **C** is much thicker than that around chamber **D**. [2]

(b) The coronary arteries supply blood to the heart muscle.
(i) Suggest **two** activities of humans which might cause a clot in a coronary artery. [2]
(ii) Explain what might be the result of such a blockage. [2]

(c) Fig 27.2 shows a plan of the circulatory system.

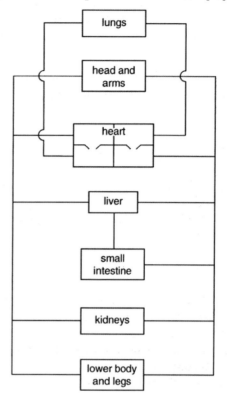

Fig 27.2

Copy Fig 27.2.
(i) Label where urea is formed. [1]
(ii) Label where urea is excreted. [1]
(iii) Show, using a series of arrows, the route taken by urea between these two organs. [2]

Total [13]

28 Fig 28.1 shows some cells from the lining of a human trachea.

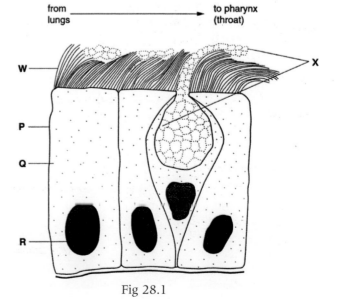

Fig 28.1

195

(a) (i) Name **P**, **Q** and **R**.
    (ii) Name the structures found in **R**.
    (iii) Identify **W** and **X**.
    (iv) Describe the functions of **W** and **X**. [5]

Smoking has a direct effect on **W** and **X**.

(b) (i) Name **three** harmful chemicals found in cigarette smoke.
    (ii) Describe the effects that these, and other substances present in cigarette smoke, may have on **W** and **X**, and also on other parts of the respiratory system. [5]

5090/2 N98   Total [10]

29 Fig 29.1 shows a food web which includes some organisms in the African grasslands.

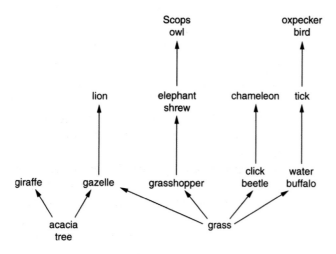

Fig 29.1

(a) (i) Draw a food chain consisting of **four** organisms. The organisms must be part of the food web. [2]
    (ii) Using examples from the food web, explain the difference between **producers** and **consumers**. [4]

(b) When weather conditions are favourable the grasshopper population can suddenly increase enormously.

Predict and explain the effect this might have on the
    (i) Scops owl population; [2]
    (ii) water buffalo population; [2]
    (iii) giraffe population. [3]

0610/2 N00   Total [13]

30 Fig 30.1 is a family tree showing the inheritance of tongue-rolling in two families. A person's ability to roll the tongue is caused by a dominant allele, **R**.

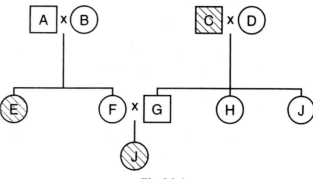

Fig 30.1

(a) Using **R** to represent the dominant allele and **r** to represent the recessive allele, what are the genotypes of individuals
    (i) A;
    (ii) F;
    (iii) J? [3]

(b) What are the chances that the next child born to F and G will be
    (i) a tongue-roller
    (ii) a female tongue-roller? [2]

5096/1 J00   Total [5]

31 Fig 31.1 shows shows a nitrogen cycle for open grassland.

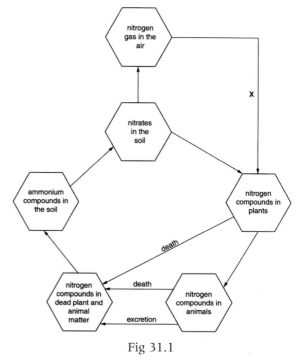

Fig 31.1

(a) (i) Name **one** nitrogen compound in plants. [1]

(ii) Name an example of a nitrogen compound which is excreted by mammals. [1]
(iii) Process **X** can only occur in certain plants. Which group of organisms carry out this process and where in a plant are they found? [2]

(b) The grassland is ploughed up and turned into farmland. Crops of maize are grown on it year after year.
(i) Predict and explain the effect of this change on the nitrogen cycle and on the crop yield. [4]
(ii) Suggest one way in which the farmer could prevent the effect on crop yield. [1]

0610/2 J00   Total [9]

32 (a) Explain, using examples, how discontinuous variation differs from continuous variation. [4]
(b) Explain, using fully-labelled genetic diagrams, how a person may inherit
(i) a blood group that shows incomplete dominance; [4]
(ii) a blood group with a recessive phenotype when both parents showed dominant phenotypes. [4]

5090/2 J01   Total [12]

33 Fig 33.1 shows the fruiting heads of five different types of cereals, **A**, **B**, **C**, **D** and **E**.

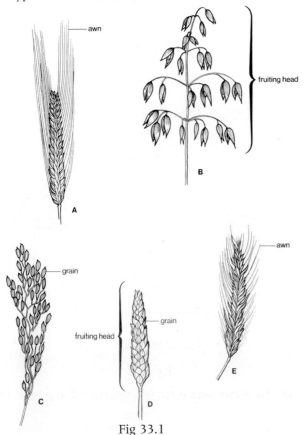

Fig 33.1

ACKNOWLEDGEMENT: S.G.Harrison et.al. The Oxford Book of Food Plants, reproduced by permission of Oxford University Press, 1969.

(a) Use the key below to identify each type of cereal. Copy Table 33.1. Write the name of each cereal in the correct box in your copy of the table.

1 Grains close together in fruiting head................. 2
  Grains not close together in fruiting head............ 4

2 Awns present... 3
  Awns absent... *Triticum*

3 Awns may be longer than fruiting head.... *Hordeum*
  Awns never longer than the fruiting head.... *Secale*

4 Grains hang down from stalks........................ *Avena*
  Grains do not hang down from stalks............ *Oryza*

Cereal	Name
A	
B	
C	
D	
E	

Table 33.1   [4]

(b) Many varieties of cereal crops grown today have been developed from wild ancestors by artificial selection.
(i) What is meant by **artificial selection**? [1]
(ii) Suggest two features which farmers might wish to introduce into a modern cereal variety. [2]

(c) Modern farming techniques often include the use of fertilisers to increase crop yield.
(i) Name the mineral ion normally included in such fertilisers which is needed by plants for the formation of all amino acids. [1]
(ii) Explain the dangers to the environment of the excessive use of these fertilisers. [4]

0610/2 N95   Total [12]

34 (a) (i) Describe the water cycle.
(ii) Explain how water may be polluted by agricultural practices. [7]
(b) Explain why the following are recycled:
(i) household water;
(ii) paper. [3]
(c) Suggest the problems which may arise when household water is recycled. [2]

5090/2 N99   Total [12]

35 (a) List the main characteristics of insects. [4]
(b) Describe how a named insect-pollinated flower is adapted to attract insects. [5]
(c) Explain the advantages and disadvantages of controlling insects with insecticides. [6]

0610/3 N98   Total [15]

36 Fig 36.1 shows the concentrations, in parts per million (p.p.m.), of an insecticide found in a lake and in organisms that rely on the lake for their existence.

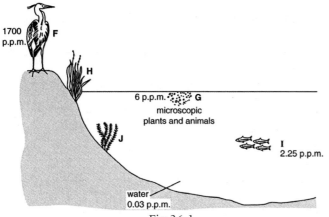

Fig 36.1

(a) Which term describes a balanced community of organisms and the habitat in which they live? [1]
(b) (i) State which organims in Fig 36.1 are producers. [1]
(ii) State the form of energy used by the producers and the form in which they pass it on to other organisms in the community. [2]
(c) (i) Draw a possible food chain for the community in Fig 36.1, using the letters shown. [1]
(ii) Explain the change in concentration of insecticide along this food chain. [2]
(d) Suggest how the insecticide may originally have entered the lake. [2]

5090/2 J01 Total [9]

37 (a) Cancer and heart disease are two major causes of death. Explain how the following may contribute towards these diseases:
(i) smoking cigarettes;
(ii) eating an unbalanced diet. [8]
(b) Suggest why smoking is now widely regarded as a socially unacceptable habit. [3]
(c) Explain how dependency on a drug, such as heroin, can lead to infection with the AIDS virus (HIV). [4]

0610/3 N95 Total [15]

38 (a) Show, by means of a table, how red and white blood cells differ in structure and function. [6]
(b) What changes are made to the blood as it flows through
(i) the lungs;
(ii) the kidneys? [4]
(c) Describe how the blood clots to seal a wound in the skin. [2]

5090/2 J00 Total [12]

39 State the difference between the terms in each of the following pairs.
(a) consumer and producer [4]
(b) carnivore and herbivore [2]

0610/2 J01 Total [6]

40 (a) With reference to sulphur dioxide as a pollutant, describe its source, effects on the environment and possible methods of control. [6]
(b) Outline the undesirable effects of deforestation. [6]
(c) Explain why non-biodegradable plastics are less environmentally friendly than biodegradable plastics. [3]

0610/3 J01 Total [15]

41 A student investigated the number of freshwater shrimps in different rivers.
The same sampling technique was used at each river. Table 41.1 shows the data collected.

name of river	oxygen concentration / arbitrary units	flow rate / m per s	temperature / °C	number of freshwater shrimps
Chesterfield	11.0	1.02	13	39
Cromford	3.0	0.005	15	0
Don	10.0	1.48	13	30
Loxley	9.5	0.56	12	28
Maun	8.0	2.64	14	0
Meden	4.0	0.01	14	3
Rivelin	10.5	0.90	13	35

Table 41.1

(a) Construct a graph to show the relationship between the number of shrimps and the oxygen concentration of the water. [5]
(b) (i) Describe the relationship between the number of shrimps and the oxygen concentration. [2]
(ii) State which river shows an unexpected result that does not fit the pattern. [1]
(iii) With reference to Table 41.1, explain this unexpected result. [2]
(c) Suggest how the student may have measured the rate of flow of the water. [3]
(d) Describe how the student could ensure that the readings of the water temperature were valid. [3]

5090/6 J01 Total [16]

Glossary

absorption: the uptake of a substance into an organism's body.

accommodation: the adjustment of the shape of the lens and eyeball, so that light is focused accurately onto the retina.

acid rain: rain containing above normal amounts of dissolved sulphur oxides and nitrogen oxides.

active site: the part of an enzyme molecule in which the substrate binds.

active transport: the movement of substances through cell membranes, against a concentration gradient, using energy.

adaptation: a feature of an organism which enables it to live successfully in its environment.

adolescence: the time between childhood and adulthood.

aerobic respiration: the release of a relatively large amount of energy from the breakdown of glucose, by combining it with oxygen.

afterbirth: the placenta, which leaves the mother's body through the vagina, just after the baby is born.

age pyramid: a type of graph showing the relative numbers of each age group of a population.

alleles: different varieties of a gene, which code for different versions of the same characteristic.

amino acids: molecules containing carbon, hydrogen, oxygen and nitrogen. A long chain of amino acids forms a protein molecule.

amnion: a membrane surrounding a developing fetus.

amniotic fluid: the liquid contained within the amnion, which supports the fetus and protects it.

amylase: an enzyme which digests starch to maltose.

anaemia: a disease caused by lack of haemoglobin, often because of a shortage of iron.

anaerobic respiration: the release of a relatively small amount of energy by the breakdown of glucose, without combining it with oxygen.

antagonistic: antagonistic muscles work in pairs, one causing the joint to bend, and the other causing it to straighten.

anther: the part of a flower where male gametes are produced.

antibiotic: a drug which kills bacteria without harming other cells.

antibodies: proteins made by white cells which attach to specific foreign cells or other substances (antigens) and help to destroy them.

antigen: a cell or other substance which is recognised as foreign by the body's white cells.

arachnid: a spider or scorpion; an arthropod with eight or more legs, which breathes by means of book lungs.

arthropod: an invertebrate with jointed legs.

artificial selection: the selection and breeding by humans of the best individuals, in order to improve the strain.

asexual reproduction: a process resulting in the production of genetically identical offspring from one parent.

assimilation: the incorporation of absorbed food into various parts of the body.

atherosclerosis: hardening of the arteries; a disease caused by deposits of cholesterol in artery walls.

atrium: one of the upper chambers of the heart, which receive blood from the veins and pass it on to the ventricles.

autotrophic nutrition: feeding by converting inorganic materials into organic ones.

auxin: a hormone produced in the meristems of plants, which affects cell elongation.

axon: a long process stretching from the cell body of a neurone, which carries impulses away from the cell body.

balanced diet: a daily intake of food containing all types of food in the correct proportions.

bicuspid valve: the valve between the left atrium and ventricle of the heart.

bile: a liquid made by the liver, stored in the gall bladder, and passed into the duodenum along the bile duct.

bile salts: substances found in bile, which emulsify fats in the duodenum.

binary fission: a form of asexual reproduction in which one cell splits into two.

blind spot: the part of the retina where the optic nerve leaves; it has no receptor cells, and so light falling onto it cannot be sensed.

bronchiole: a small tube carrying air to and from the alveoli in the lungs.

bronchitis: an infection in the bronchi, causing coughing.

bronchus: a large tube connecting the trachea to the bronchioles.

cancer: a disease resulting from the uncontrolled division of cells in one or more parts of the body.

carbohydrase: an enzyme which breaks down carbohydrate molecules.

carbohydrate: sugars and starches.

carnivore: an animal which feeds on other animals, which it kills.

carrier: a heterozygous organism, possessing a recessive allele which does not show in its phenotype.

cartilage: a tissue made of living cells in a matrix of collagen. It is found in several places in the mammalian skeleton, and makes up the entire skeleton of sharks.

catalyst: a substance which alters the rate of a reaction, without being changed itself.

cell sap: a solution of sugars, amino acids and many other substances, found in the vacuoles of plant cells.

cell surface membrane: a very thin layer of protein and fat, which surrounds the cytoplasm of every living cell.

cellulose: a polysaccharide (long-chain carbohydrate) made from glucose molecules linked together in very long chains. Cellulose forms fibres, which make up the cell walls of plants.

cerebellum: an area of the brain which controls muscular coordination.

cerebrum: the part of the brain responsible for conscious thought, language and personality. In mammals it is very large and folded, and forms two cerebral hemispheres.

chlorophyll: a green pigment found in all plants and some bacteria and protoctists, which absorbs energy from sunlight to be used in photosynthesis.

chloroplasts: organelles found in many plant cells, which contain chlorophyll, and where photosynthesis takes place.

choroid: a black layer lining the eye, which absorbs light and so cuts down reflections inside the eye.

chromosome: a coiled thread of DNA and protein, found in the nucleus of cells.

chyme: a mixture of partly digested food, enzymes and hydrochloric acid, resulting from digestion in the stomach.

cilia: small hair-like structures which project from some cells, and perform waving movements in synchrony with each other.

ciliary muscle: a ring of muscle surrounding the lens in the eye, which can adjust the size of the lens.

clone: a group of genetically identical organisms.

codominance: the existence of two alleles for a characteristic where neither is dominant over the other.

colon: the part of the alimentary canal between the ileum and the rectum, where water is absorbed.

community: all the organisms which live in a particular habitat.

conservation: maintaining the environment in a state in which natural wildlife can flourish.

consumer: an organism which consumes other organisms for food; all animals are consumers.

continuous variation: variation where variants do not belong to definite categories, but may fit anywhere within a wide range.

corpus luteum: a structure in a mammalian ovary, formed from a follicle, which secretes the hormone progesterone.

cortex: the part of a stem or root between and around the vascular bundles or stele.

cuticle: a waxy, waterproof covering, found on outer surfaces of a leaf.

deamination: a reaction which takes place in the liver, where amino acids are converted to urea and carbohydrate.

deforestation: the removal of trees.

denitrifying bacteria: bacteria which often live in damp soil, and which convert nitrates into nitrogen gas.

destarching: keeping a plant in the dark, so it cannot photosynthesise and will use up its starch stores.

diabetes: a disease caused by insufficient secretion of insulin by the pancreas, resulting in widely fluctuating blood glucose levels.

dialysis: the separation of a mixture of dissolved solutes by diffusion through a partially permeable membrane.

dichotomous key: a way of identifying an organism, in which you are given successive pairs of descriptions to choose between.

diffusion: the movement of particles of gas, solvent or solute, from an area of high concentration to an area of low concentration; that is, down a concentration gradient.

diploid cell: a cell which has two sets of chromosomes.

disaccharide: a sugar made of molecules which consist of two monosaccharide units joined together.

discontinuous variation: variation where the variants belong to distinct categories with no intermediates.

DNA: deoxyribonucleic acid; a substance found in chromosomes, which carries a code used by the cell when making proteins.

dominant allele: an allele which has the same effect on the phenotype of an organism, whether the organism is homozygous or heterozygous for that gene.

Down's syndrome: a condition caused by having an extra chromosome.

drug: an externally administered substance which modifies or affects chemical reactions in the body.

duodenum: a short length of alimentary canal into which the bile duct and the pancreatic duct empty bile and pancreatic juice.

ecosystem: the living organisms and their environment, in a certain area e.g. a wood, or a pond.

effector: a part of an organism which carries out an action, often in response to a stimulus, e.g. muscles and glands.

egestion: the removal of indigestible food from the body.

embryo: a plant or animal as it develops from a fertilised egg.

emulsification: the breaking up of large droplets of fat into small ones.

enzymes: biological catalysts; proteins made by living organisms, which speed up chemical reactions.

eutrophication: a process in which extra nutrients are added to water, so increasing the growth of plants and bacteria, and reducing oxygen levels.

excretion: the removal of toxic materials and the waste products of metabolism, from organisms.

extensor: a muscle which straightens a limb when it contracts.

extracellular: outside cells.

faeces: remains of indigestible food, bacteria, mucus etc., which are egested from the alimentary canal.

family: a group of genera with similar characteristics.

fat, saturated: the type of fat found in many animal products such as meat, milk, butter and chocolate.

fat, unsaturated: the type of fat found in many plant products such as olive oil, sunflower oil.

fatty acids: molecules which can combine with glycerol to make fats.

fermenter: a vessel in which microorganisms such as fungi or bacteria are grown under controlled conditions, usually on a large scale, in order to produce substances such as alcohol, antibiotics or food substances.

fertilisation: the joining together of the nucleus of a male gamete with the nucleus of a female gamete.

fetus: a mammalian embryo in a fairly advanced stage of development.

flaccid: a flaccid plant cell is one which has lost water, so that the cytoplasm does not push outward on the cell wall.

flexor: a muscle which bends a limb when it contracts.

fovea: the part of the retina where receptor cells are most densely packed, and onto which light is normally focused.

fruit: a plant's ovary after fertilisation; it contains seeds, and usually helps in their dispersal.

gamete: a sex cell, containing only one set of chromosomes (i.e. haploid). Eggs and sperm are gametes.

gene: a length of DNA which codes for the making of a particular protein; a unit of inheritance.

genetic engineering: the manipulation of genetic material to produce new types of organisms.

genotype: the genes possessed by an organism.

genus: a group of species with similar characteristics.

geotropism: the directional response of a plant, by growth, to gravity.

glucagon: a hormone secreted by the islets of Langerhans in the pancreas in response to low concentrations of glucose in the blood. It stimulates the liver to release glucose from its glycogen stores, so raising blood glucose concentration.

glycogen: a polysaccharide used as a storage substance in the cells of many animals and fungi; in humans, it is stored in liver and muscle cells.

greenhouse effect: the trapping of heat by a layer of gases in the Earth's atmosphere, especially carbon dioxide, water vapour and methane. Without this effect, the Earth would be too cold to sustain life.

habitat: the place where an organism lives.
haemoglobin: a protein containing iron, found in red blood cells, which carries oxygen.
haploid: having only one set of chromosomes.
herbivore: an animal which eats plants.
hermaphrodite: able to make both male and female gametes.
heterotrophic: using food made by other organisms; all animals and fungi are heterotrophic.
heterozygous: possessing two different alleles.
homeostasis: the maintenance of a constant internal environment.
homeothermic: able to maintain a constant body temperature.
homologous chromosomes: chromosomes which carry genes for the same characteristics in the same positions.
homozygous: possessing two identical alleles.
hormone: In animals, hormones are usually made in endocrine glands and transported in the blood; the organs whose activity they affect are called target organs.
hypothalamus: a part of the brain which is responsible for coordinating temperature regulation.
ileum: the part of the alimentary canal between the duodenum and colon; it is very long, and is lined with villi to help with absorption of digested food.
immunity: the possession of antibodies against a particular disease.
immunosuppressants: drugs which stop a person's immune system from responding to antigens. They are used to prevent the rejection of a transplanted organ.
incisor: a tooth at the front of a mammal's mouth, normally used for biting off pieces of food for chewing.
ingestion: taking food into the alimentary canal.
insecticide: a chemical which kills insects.
insecticide, systemic: an insecticide which is taken into a plant's vascular bundles, and so is carried to all parts of it.
islets of Langerhans: patches of cells in the pancreas which secrete insulin and glucagon.
keratin: the protein which makes up hair, nail, horn and the outer layer of skin.
kwashiorkor: a type of malnutrition caused by a lack of protein in the diet.
lactation: the secretion of milk by a female mammal.
lacteal: a lymphatic vessel inside a villus, which looks milky because it contains absorbed fat.
lactic acid: a chemical produced during anaerobic respiration in animals.
lactose: milk sugar; a disaccharide found in milk.
lamina: the blade of a leaf.
leaching: the loss of soluble substances from soil, as they are washed out by rain water.
limiting factor: a factor whose supply limits the rate of a metabolic reaction; e.g. low light intensity may limit the rate at which photosynthesis takes place.
lipase: an enzyme which digests fats.
lumen: the space in the middle of a tube.
lymph node: an organ through which lymph flows, containing many white cells, and where antibodies are made.
malnutrition: a condition caused by eating an unbalanced diet, especially if the diet is badly lacking in one or more types of food.

maltase: an enzyme which digests the disaccharide maltose.
maltose: a disaccharide found in germinating seeds, formed from the breakdown of starch.
mechanical digestion: the breakdown of large particles of food into small ones, by teeth and the churning movements produced by muscles.
medulla oblongata: the part of the brain nearest to the spinal cord, responsible for the control of heart beat and breathing movements.
meiosis: a type of cell division in which four haploid cells, genetically different from each other, are produced from one diploid cell.
menstruation: the breakdown and loss of the soft lining of the uterus.
mesophyll: the central layers of a leaf, where photosynthesis takes place.
metabolism: the chemical reactions taking place in a living organism.
micropyle: a small gap in the integuments of an ovule, through which the pollen tube grows. Later, when the ovule becomes a seed, the micropyle remains as a hole through which water enters at germination.
mitosis: a type of cell division in which two genetically identical cells are formed from a parent cell.
mitral valve: the valve between the left atrium and ventricle.
molar: a large tooth near the back of a mammal's mouth, used for chewing, grinding or slicing.
monosaccharide: a simple sugar; a sugar whose molecules are made of a single sugar unit.
motor neurone: a nerve cell which carries impulses from the CNS to an effector.
mucus: a slimy liquid, secreted by goblet cells, used in many parts of the body for lubrication.
mutation: an unpredictable change in an organism's genes or chromosomes.
mycelium: the tangle of threads (hyphae) which makes up the body of a fungus.
natural selection: the selection of only the best adapted organisms for survival and reproduction, by natural factors such as predators or shortage of food supply.
nectar: a sugary liquid secreted by flowers to attract insects or birds for pollination.
nerve: a group of nerve fibres, surrounded by connective tissue.
nerve fibre: an axon or dendron; a strand of cytoplasm extending from a nerve cell body.
neurone: a nerve cell
niche: the role of a living organism in a community.
nitrifying bacteria: bacteria which convert proteins and urea into nitrates.
nitrogen fixation: the conversion of nitrogen gas into a compound of nitrogen, such as ammonia, nitrates or proteins.
obesity: being considerably overweight.
oesophagus: the part of the alimentary canal between the mouth and the stomach.
oestrogen: a hormone secreted by the ovaries which produces female secondary sexual characteristics.
organelle: a structure inside a cell.
organ: part of an organism; a structure made of several tissues which performs a particular function, e.g. heart.
osmoregulation: the control of the water content of the body.
osmosis: the diffusion of water molecules from a dilute solution to a concentrated solution, through a partially permeable membrane.

ovulation: the release of an egg from the ovary.

ovule: a structure inside a plant's ovary which contains a female gamete, and which develops into a seed after fertilisation.

ovum: a female gamete; an egg.

oxyhaemoglobin: haemoglobin combined with oxygen; it is a brighter red than haemoglobin.

palisade layer: a layer of rectangular cells near the upper surface of a leaf, where most photosynthesis takes place.

pancreatic juice: watery fluid secreted by the pancreas, containing various digestive enzymes, which flows into the duodenum along the pancreatic duct.

pathogen: an organism which causes disease.

pepsin: an enzyme secreted by glands in the wall of the stomach, which digests proteins.

pericarp: the outer layers of a fruit, developed from the ovary.

peristalsis: rhythmic contractions of the muscles in the walls of tubes, such as the alimentary canal or oviduct, which squeeze the contents along.

permeable: allowing substances to pass through.

pesticides: chemicals used to kill pests.

pH: a measure of the acidity of a solution; pH 7 is neutral, below 7 acidic, and above 7 alkaline.

phagocytosis: 'cell feeding'; the intake of particles of food by a cell.

phenotype: characteristics shown by an organism, a result of interaction between its genotype and its environment.

phloem: a plant tissue in which substances made by the plant are carried from one part to another.

phylum: a major group of organisms; a subdivision of a kingdom.

placenta: the organ through which a mammalian embryo is connected to its mother, and through which it obtains food, oxygen etc.

plaque: a mixture of food remains and bacteria which builds up on and between teeth.

plasma: the liquid part of blood.

plasmolysis: shrinkage of the cytoplasm of a plant cell, so that the cell membrane begins to tear away from the cell wall; caused by loss of water.

pleural membranes: membranes surrounding the lungs and lining the thoracic cavity.

plumule: part of an embryo plant which will develop into the shoot.

pollination: the transfer of pollen from an anther to the stigma.

pollution: the addition of substances to the environment which harm life.

polysaccharide: a carbohydrate such as starch or cellulose, whose molecules are made of many sugar units joined together.

population: all the organisms of a particular species living in a certain area.

predator: an animal which hunts and kills other animals (known as its prey) for food.

premolar: large tooth between the canines and molars of mammals; unlike molars, premolars are present in the milk dentition as well as the permanent dentition.

primary consumer: the first consumer in a food chain; a herbivore.

producer: the first organism in a food chain, which produces food, i.e. a green plant.

progesterone: a hormone secreted by the ovary and later the placenta, which maintains the uterus lining during pregnancy.

prostate gland: a gland near the junction of the two sperm ducts with the urethra, which secretes a fluid in which sperm swim.

protease: an enzyme which digests protein.

protein: a substance whose molecules are made of long chains of amino acids.

puberty: the age at which secondary sexual characteristics appear, and gametes begin to be produced.

pure breeding: producing offspring like themselves; homozygous.

radicle: a young root.

receptor: part of an organism which receives stimuli.

recessive allele: an allele which only shows in the phenotype when no dominant allele is present.

rectum: the last part of the alimentary canal, in which faeces are formed before being egested through the anus.

reflex action: an automatic, fast response to a stimulus.

reflex arc: the series of neurones and synapses by which an impulse passes from a receptor to an effector in a reflex action.

refraction: the bending of light rays as they pass through materials of different densities.

relay neurone: a neurone in the central nervous system, which passes impulses from one neurone to another.

respiration: the release of energy from food substances such as carbohydrates; it happens in every living cell.

rickets: a disease of the bones caused by lack of vitamin D.

root hair: part of a root cell which projects into the soil, where it absorbs water and mineral salts.

roughage: fibrous, indigestible food, which stimulates the muscles of the alimentary canal to perform peristalsis.

saliva: watery fluid containing amylase and mucus, secreted into the mouth by salivary glands.

saprophyte: an organism which feeds on dead organic material, by secreting enzymes onto it and absorbing it in liquid form.

sclera: the tough outer coat of the eyeball.

secretion: the production and release of a useful substance.

selection pressure: a factor acting on a population which favours certain varieties for survival.

sensory neurone: a nerve cell which carries impulses from a receptor to the CNS.

sex chromosomes: chromosomes which determine the sex of an organism.

sexual reproduction: a process involving the fusion of nuclei to form a zygote and the production of genetically dissimilar offspring.

single cell protein: high protein food made from microorganisms, such as mycoprotein made from fungi.

solute: a substance which dissolves in water or another solvent, to form a solution.

solvent: a liquid such as water, in which other substances can dissolve.

species: a group of organisms with similar characteristics, which can breed with each other, but not with organisms of different species.

spermatozoa: sperm; the male gametes of animals.

sphincter muscle: a muscle round a tube, which can close the tube when it contracts.

sphygmomanometer: an instrument used to measure blood pressure.

spinal nerve: a nerve entering or leaving the spinal cord.

spongy mesophyll: layer of cells near the underside of a leaf where photosynthesis takes place; they have large air spaces between them.

starch: the polysaccharide storage material of plants, made from molecules of hundreds of glucose units linked together.

stimulus: a change in an organism's environment which is detected by a receptor.

substrate: (a) a substance which is converted to another during a chemical reaction, (b) the material on which a bacterium or fungus lives and feeds.

sucrase: an enzyme which digests sucrose.

sucrose: a non-reducing disaccharide sugar.

sugar: a type of carbohydrate made of molecules consisting of one (monosaccharide) or two (disaccharide) sugar units; it tastes sweet, and is soluble.

sweat gland: coiled gland in the dermis, which extracts water, salt and urea from the blood and secretes it as sweat.

synapse: very small gap between two nerve cells.

synovial joint: a joint between two bones, where free movement can occur.

synovial membrane: a membrane enclosing a synovial joint, attached to the bones on each side, which secretes and encloses synovial fluid.

tendon: tough band of fibres which joins a muscle to a bone.

testa: hard outer covering of a seed.

test cross: breeding an organism showing a dominant characteristic in its phenotype, with one which is homozygous for the recessive characteristic. The offspring from this cross will show you whether the first organism is homozygous or heterozygous.

testosterone: a hormone secreted by the testes, responsible for male secondary sexual characteristics.

thorax: chest; the part of the body containing heart and lungs, separated from the abdomen by the diaphragm.

thrombosis: a blood clot in a vein or artery.

tissue: a group of similar cells which together perform a particular function.

tissue fluid: fluid which fills in spaces between cells in the body, formed by plasma which leaks from blood capillaries.

toxin: a poison, especially one produced by pathogens inside the body.

trachea: a strengthened tube in vertebrates or insects, through which air passes on its way to and from the respiratory surface.

translocation: the movement of materials within a plant, particularly ones which the plant itself has made, such as sugars.

transpiration: the loss of water vapour from a plant, mostly from the leaves.

tricuspid valve: the valve between the right atrium and ventricle.

trophic level: the level in a food chain at which an organism feeds.

tropism: a directional growth response by a plant to a stimulus.

turgid: a turgid plant cell contains plenty of water, so that the cytoplasm pushes outwards on the cell wall.

trypsin: an enzyme secreted by the pancreas, which digests proteins in the duodenum.

umbilical cord: cord containing an artery and vein, which connects a fetus to its placenta.

urea: substance containing nitrogen, made in the liver by the deamination of excess amino acids, and excreted in urine.

ureter: a tube carrying urine from the kidney to the bladder.

urethra: a tube leading from the bladder to the outside; it carries urine, and also semen in males.

urine: a watery liquid containing urea and other excretory substances, produced by the kidneys.

uterus: womb; muscular organ with a soft lining, where a fetus develops.

vacuole: an organelle containing liquid, and surrounded by a membrane.

vagina: tube leading from the uterus to the outside.

vascular bundle: a group of xylem vessels and phloem tubes.

vas deferens: sperm duct; the tube carrying sperm from the testis to the urethra.

vasoconstriction: a decrease in the diameter of blood vessels, caused by contraction of muscle in the walls of arterioles.

vasodilation: an increase in the diameter of blood vessels, caused by the relaxation of muscle in the walls of arterioles.

vector: an organism, or part of an organism, which transfers something from one kind of organism to another. Plasmids and viruses can be used as vectors to transfer DNA from one cell to another in genetic engineering. Mosquitoes are vectors for malaria, transferring *Plasmodium* from one person to another.

ventricle: one of the thick-walled lower chambers of the heart, which pumps blood into the arteries.

villus: a small 'finger' of tissue, which increases surface area; found e.g. in the ileum and the placenta.

xylem vessel: a long, narrow tube made of many dead, lignified cells arranged end to end; conducts water in plants, and supports them.

yolk: a store of fat and protein food in an egg.

zygote: a cell formed at fertilisation, which will develop into an embryo.

Index

absorption, in small intestine 31, 32
accommodation 92–93
acid rain 173–174
activated sludge 150–152
active site 12–13
active transport 10, 70
addiction, to nicotine 53
  to heroin 145
adipose tissue 20
adolescence 80
adrenal glands 100–101
adrenaline 101
aerobic exercise 143
afterbirth 78
agar jelly 148
age pyramid 167
AIDS 141
alcohol 144–145, 153
alimentary canal 29–33
alleles 120
alveoli 46–49
amino acids 19
amnion 78
amniotic fluid 78
amphibians 185
amylase 12, 14, 15, 25, 30, 87, 153
anaemia 21
androgens 80
animals, classification 183–185
annelids 183
Anopheles 136–138
antagonistic muscles 118
ante-natal care 79
anthers 82–83
antibiotics 136, 142, 155–156
  resistance to 131
antibodies 139–141
antigens 140–141
appendix 29
arachnids 184
arteries 57, 58, 59
arterioles 57, 108
arthropods 184
artificial insemination 82
artificial selection 132
asexual reproduction 71–73, 89
atherosclerosis 143
atria 54–57

autotrophic nutrition 34
auxin 103
axon 95–96
B cells (lymphocytes) 139, 140
bacteria, characteristics of 182
  denitrifying 164
  in sewage treatment 150–152
  in yoghurt and cheese 154
  nitrifying 164
  nitrogen-fixing 164
  reproduction 72
balanced diet 22–23
behaviour 98, 99
Benedict's solution 18
biceps 118
bile 25, 30
bile duct 29
binary fission 72
biodiversity 179
biological control 138
biological washing powder 15
biotechnology 147–158
birds 185
birth 78
birth control 81
Biston betularia 130–131
biuret test 19
bladder 113–114
blind spot 92
blood 59–62
  cells 59–60
  clotting 61
  group, inheritance of 122, 125
  plasma 60, 61–62
  vessels 54, 57–59
bone 116
brain 96, 98–99
bread 153
breathing 49–50, 52
bronchus 47–48
BST 102
cacti 69
caecum 29
calcium 21
callus 73
cancer 53
canines 26–27
capillaries 57, 58

capillary network   63
carbohydrates   17–18, 20
    digestion of   25
carbon cycle   163
carbon dioxide   36, 41
    and global warming   171–172
    transport in blood   61
carbon monoxide   53
cardiac arrest   55
cardiac muscle   54
carnivorous plants   164
carpel   82
carrier   121
carrier proteins   10
cartilage   47, 116, 117
catalase   12
catalysts   12
cell membrane   2
cell sap   2
cell structure   1–3
cell surface membrane   2
cell wall   2
cells, animal and plant compared   5
cellulose   2, 17
cement, of tooth   26
central nervous system   96, 98
centromere   120
cerebellum   98, 99
cerebrum   98, 99
cervix   74, 75, 77
CHD   143
cheese   154
chemical digestion   33
chlorophyll   2, 34, 37
chloroplasts   2
cholesterol   22
chordates   185
choroid   92, 93
chromatid   120
chromosomes   3, 71, 73, 74, 119–126
chyme   30
cilia   47, 53, 76
ciliary muscle   92, 93–94
circulatory system   54–58
cirrhosis   144
classification   181–185
clear felling   179
clinostat   104
clone   71
codominance   121

cold virus   141
collagen   116
colon   29, 31
community   159
companion cell   64
condom   81
cones   92
conjunctiva   92
conservation   179–180
constipation   21
consumers   160–161
continuous variation   127
control   39
coppicing   179
cornea   92–93
cornified layer   107
coronary arteries   55, 56
coronary heart disease   143
corpus luteum   101
cotyledons   87
cross-pollination   84
crown, of tooth   26
crustaceans   184
cuttings   73
cystic fibrosis   120–121, 158
cytoplasm   2
Darwin   129
DDT   178–179
deamination   112
decomposers   163
deficiency disease   21
deforestation   175–176
    and global warming   172
dendron   95–96
denitrifying bacteria   164
dentine   26
depressant   144, 145
destarching   39
development   88
diabetes (mellitus)   111
dialysis   114
diaphragm   29, 47, 49
diastole   56
dichotomous key   169
dicotyledonous plants   183
diet   17–24
diffusion   6
digestion, in humans   24–33
digestion, in *Mucor*   72
diploid cells   74

disaccharides 17
discontinuous variation 127
disease 134–145
dispersal, of fruits and seeds 86–87
DNA 71, 119
dominant alleles 121
dormancy 87
dry mass 88
duodenum 29, 30–33
ecosystem 159
effectors 95
effluent 150
egg 73
ejaculation 76
elbow 117, 118
embryo 77
   in seed 86, 87
emphysema 53
emulsification 31
emulsion test 20
enamel 26
endocrine glands 100
energy, from carbohydrates 18
   from food 22
   in food chains 160–161
environment, human effects 170–180
enzymes 12–16, 87
   in digestion 25
   in biological washing powders 15
   in the food industry 15
epidermis, of leaf 35–36
epiglottis 29, 47
erector muscles 108
ethanol test 20
ethene 103
eutrophication 150, 176–177
evolution 132
excretion 112–114
exercise, benefits of 143
expiration 50
extensor muscle 118
eye 92–95
F1 and F2 generations 125
fats 20, 108
   digestion of 25
   saturated 22
   unsaturated 23
fatty acids 20
fermentation 153
fermenter 156

fertilisation 73, 76–77
fertilisation, in flower 84–85
fertilisers, and pollution 176–177
fertility drugs 82
fetus 77
fibrin 61
fibrinogen 61
filament 82–83
fish 185
fisheries 180
flexor muscle 118
flowers 82–86
fluoride 28
focusing 93–94
food additives 23, 24
food chains 160–161
food industry, uses of enzymes in 15
food web 162
fossil fuels, and acid rain 173–174
   and carbon cycle 163
   and global warming 172
fovea 92
fructose 15
fruits 86–87
   ripening 103
fungi 182
   in cheese 154
   reproduction 72
fungicides 178
Fusarium 155
gall bladder 29
gametes 73, 83
gas exchange surface 46
gaseous exchange 46, 52
gastric juice 30
gene therapy 158
genetic diagram 123
genetic engineering 156–158
genetics 119–126
genotype 121
genus 181
geotropism 102–103
germination 87, 88
gestation period 78
glasshouse, cultivation in 41–42
global warming 171–173
glucagon 100, 110–111
glucose 15, 17, 18, 34, 36–37
   in respiration 4
   regulation of 110–111

gluten   153
glycerol   20
glycogen   110–111
goblet cells   47
gonorrhoea   136
greenhouse effect   171
growth   88
guard cells   35, 36, 69
gum disease   27
habitat   159
haemoglobin   53, 59, 61
haemophilia   157
hairs   107–108
haploid cells   74
heart   54–57
heart attack   55, 143
heart beat   56
heart disease   22–23
heavy metals   178
hepatic artery   58–59
hepatic portal vein   32
hepatic vein   58–59
herbaceous plant   118
herbicides   178
heroin   144–145
heterozygous   120
hilum   86–87
HIV   141
homeostasis   62, 106–111
homeothermic animals   106
homologous chromosomes   120
homozygous   120
hormones, in animals   95, 100–102
   in plants   103
   sex   80, 81, 82
human growth hormone   157
human population   168
humerus   117, 118
hydrochloric acid   30
hydrogencarbonate indicator   51
hyphae   72
hypothalamus   98, 99, 107
ileum   29
immune system   140, 142
immunity   140
immunosupressants   142
implantation   77
incisors   26–27
incubation time   136
influenza   135, 140

inheritance   119–126
insect pollination   83, 86
insecticides   138
insects   184
inspiration   49–50
inspired and expired air   50
insulin   100, 110–111, 156
intercostal muscles   47, 49–50
interphase   71
involuntary actions   99
iodine solution   18, 39
iris   92, 95
iron   21
islets of Langerhans   110
isomerase   15
joints   117
keratin   107
keys   160
kidney machine   114
kidney transplants   114
kidneys   113–114
kwashiorkor   23
lactase   31
lactation   79
lactic acid   51
Lactobacillus   154
lamina, of leaf   35
large intestine   29, 31–32
larynx   47
leaves   34–36, 39
lens   92, 93–94
ligaments   117
lignin   64
limiting factors, in photosynthesis   40
   for population growth   166–167
Linnaeus   181
lipase   15, 25, 30, 31
liver   29, 32, 110–111, 112–113
lungs   46–49
lymph   62–63
lymph nodes   60, 63
lymphatic vessels   63
lymphocytes   139–141
magnesium   37, 39
malaria   132, 136–138
malnutrition   23
Malpighian layer   107
maltase   31
maltose   17
mammals   185

medulla oblongata   98, 99
meiosis   74
melanism   130–131
memory cells   140
meninges   98
menstrual cycle   80, 101
mesophyll   35, 36
metabolic reactions   12
methane   152, 171, 173
micropyle   84–85
milk   79
milk teeth   27
minerals, in animal nutrition   21
   uptake by plants   70
mitochondria   2
mitosis   71
molars   26–27
molluscs   183
monocotyledonous plants   183
monosaccharides   17
mosquitoes   136–138
motor neurone   96, 97
Mucor   72
mucus   29, 47, 53, 120
muscle   117, 118
mycelium   72
mycoprotein   155
myelination   96
myriapods   184
narcotic   144
natural selection   129
nectary   82
negative feedback   109
nematodes   184
nervous system   95–99
neurones   95–98
niche   159
nicotine   52–53
nitrate   37, 39
nitrifying bacteria   164
nitrogen cycle   163–164
nitrogen fixation   163–164
normal distribution   128
nucleus   2, 3
nutrition, of animals   17–33
   of plants   34–42
obesity   23
oesophagus   29, 30
oestrogen   80, 100, 101
optic nerve   92–93

optimum temperature (for enzymes)   13
organelle   2, 4
organs   4
osmosis   6–9, 10, 11, 65–66
ovary, in humans   74, 75, 77
   in flower   82–83
oviduct   74, 75, 76
ovulation   75
ovule   82
oxidation   44
oxygen debt   51
oxygen, transport in blood   61
pacemaker   56
palate   29, 46
palisade layer   35
pancreas   29, 30, 100, 110–111
partially permeable membrane   7
pathogens   135
pectinase   15, 16
penicillin   131, 142, 155
penis   75, 76
peppered moths   130–131
pepsin   30
peristalsis   29–30, 76
pesticides   178–179
   contact   70
   systemic   70
petals   82
pH, effect on enzymes   13, 14
phagocytosis   139
phenotype   121
phloem   64, 70
phloem tubes   35
photosynthesis   34–42
   equation   34
   factors affecting   40–42
phototropism   102–103
placenta   77–78
plant hormones   73
plants, classification   183
plaque   27
plasma membrane   2
plasmid   156
Plasmodium   136–137
plasmolysis   9
platelets   59, 60–61
pleural membranes   47
plumule   87
poikilothermic animals   106
pollen   82–83

pollen sacs   83
pollen tube   84–85
pollination   83–84, 86
polysaccharides   17
populations   159, 165–168
   human   168
potato, reproduction   72–73
potometer   67
pregnancy   79
premolars   26–27
primary consumers   160–161
producers   160–161
product   12
progesterone   100, 101–102
prostate gland   75
protease   12, 15, 25, 30
proteins   19, 20
   digestion of   25
puberty   80
pulmonary embolism   58
pulmonary vessels   55
pulp cavity   26
pupil   92, 95
pyramids, of biomass   162
   of energy   162
   of numbers   161
radicle   87
receptors   92
recessive alleles   121
rectum   29, 32
recycling   180
red blood cells   59–60
reducing sugars, testing for   18
reduction division   74
reflex action   97
reflex arc   97
rejection, of transplant   142
relay neurone   97
reproduction, asexual   71–73
   in flowering plants   82–88
   in humans   74–82
   sexual   73–88
reptiles   185
respiration   44–53
   aerobic   44
   anaerobic   45, 51
   equation   44
retina   92–95
Rhizobium   164
ribs   49

rickets   21
rods   92
root cap   65
root hairs   65–66, 70
root, of tooth   26
root, structure   64–65
roughage   21
rubella   79
salivary gland   29, 30
saprotrophic nutrition   152
sclera   92
scrotum   75
scurvy   21
sebaceous glands   107
secondary consumers   160–161
seeds   86–87
selection pressure   131
selective felling   179
self-pollination   84
semen   76
seminal vesicles   75
sense organs   92
sensory neurone   97
sepals   82
sewage, and pollution   176–177
   treatment   150
sex chromosomes   125
sex inheritance   125
sexual reproduction   89
sexually transmitted diseases   81
shivering   108
sickle cell anaemia   131–132, 134
sieve plate   64
single cell protein   155
skeletal muscle   118
skeleton   116–117
skin   107–109
skull, human   26
small intestine   29, 30–31, 32, 33
smoking   52–53
species   181
species diversity   175
sperm   73, 75–76
sperm ducts   75
sphincter muscle   29
spinal cord   96, 97
spongy layer   35
sporangium   72
spores   72
stabilising selection   132

stamens 82–83
starch 17–18, 37
   testing for 18
   testing for in leaves 39–40
stem, structure 65
sterilisation 81
sternum 49
stigma 82
stimuli 92
stomach 29, 30
stomata 35, 41, 67, 69
stroke 143
style 82
substrate 12
sucrase 31
sucrose 17, 38
sugars, testing for 18
sulphur dioxide 174
suspensory ligaments 92, 93–94
sweat glands 107, 109
synapse 97
synovial joint 117
syphilis 136
systole 56
T cells (lymphocytes) 140
tar, in cigarettes 53
taxis 99
tear glands 92
teeth 26–28
temperature control 107–109
temperature, effect on enzymes 13–14
test cross 124
testa 87
testes 75
testosterone 80, 100
thrombosis 58
tissue culture 73
tissue fluid 62–63
tissues 3–4
tooth decay 27–28
toxins 135
trachea 47
translocation 70
transmitter substance 97
transpiration 67–70
transplants 142
transport, in mammals 54–63
   in plants 64–70
triceps 118
trophic levels 161

tropisms 102–103
trypsin 30
tubers, of potato 73
turgidity 9, 118
ulna 117, 118
umbilical cord 77–78
urea 112–114
ureter 113
urethra 74–75
urine 113–114
uterus 74, 75, 76, 77, 78
vacuoles 2
vagina 74, 75, 76
valves, in heart 55, 56–57
   in veins 57, 58
variation 126–128
vascular bundle 64–65
vasoconstriction 108
vasodilation 108
vector 136
vein, of leaf 35
veins 57, 58, 59
vena cava 55
ventilation 49–50
ventricles 54–57
venules 57
vertebrates 185
villi 31
viruses 135
   life cycle 148
vitamins 20–21
voluntary actions 99
water cycle 165
water potential 7, 67
weedkillers 103
wet mass 88
white blood cells 59, 139–142
wilting 69, 118
wind pollination 84, 86
world food supplies 23
xylem 64–66, 118
xylem vessels 35, 36
yeast 152–153
   population growth 166
yoghurt 154
yolk 75
zygote 73